T0039150

SPACE EXPLORATION

A History *in*
100 Objects

ALSO BY STEN ODENWALD

SPACE EXPLORATION

A History *in* 100 Objects

STEN ODENWALD

Foreword by John Mather

THE EXPERIMENT

NEW YORK

The Experiment, LLC
220 East 23rd Street, Suite 600
New York, NY 10010-4658
theexperimentpublishing.com

THE EXPERIMENT and its colophon are registered trademarks of The Experiment, LLC. Many of the designations used by manufacturers and sellers to distinguish their products are claimed as trademarks. Where those designations appear in this book and The Experiment was aware of a trademark claim, the designations have been capitalized.

The Experiment's books are available at special discounts when purchased in bulk for premiums and sales promotions as well as for fund-raising or educational use. For details, contact us at info@theexperimentpublishing.com.

Library of Congress Cataloging-in-Publication Data

Names: Odenwald, Sten F., author.
Title: Space exploration : a history in 100 objects / Sten Odenwald.
Description: New York : The Experiment [2019]
Identifiers: LCCN 2019026580 (print) | LCCN 2019026581 (ebook) | ISBN
    9781615196142 (hardcover) | ISBN 9781615196159 (ebook)
Subjects: LCSH: Astronomy--History. | Astronautics--History. | Outer
    space--Exploration--History.
Classification: LCC QB500.262 .O34 2019 (print) | LCC QB500.262 (ebook) |
    DDC 520.9--dc23
LC record available at https://lccn.loc.gov/2019026580
LC ebook record available at https://lccn.loc.gov/2019026581

ISBN 978-1-61519-614-2
Ebook ISBN 978-1-61519-615-9

Cover and text design by Beth Bugler and Jack Dunnington
Cover photographs courtesy of NASA (galaxy) and the Smithsonian National Air and Space Museum
    (Sputnik 1; photo by Eric Long [NASM 2006-25353])

Manufactured in China

First printing November 2019
10 9 8 7 6 5 4 3 2 1

To my wife, Susan, and my daughters, Emily and Stacia

CONTENTS

Space Exploration—A History in 100 Objects is filled with fascinating stories, and you can certainly enjoy them in any order. But if you want to take in the full measure of human ingenuity with regard to our understanding of space, it's better to go from beginning to end. Sten Odenwald has surprises for you on every page—beginning with the very first entry, on a seemingly simple piece of rock dating back many thousands of years; it may not look like much, but it paves the way for all the momentous breakthroughs that follow.

Every single essay in this collection of objects is a brilliant and fun read. Together, they tell a breathtaking story that starts with early humans writing their calendars, surveying their fields, and then—only a few thousand years later—living everywhere, exploring everything; building telescopes and prying open the universe's secrets. Odenwald doesn't just describe the objects; he weaves the history of our species with our growing body of knowledge about these objects. Here, you'll see the tools of the astronomer's trade: star charts and celestial catalogs, calculators and maps, telescopes and satellites, and robotic explorers of the solar system. But you'll also see objects that are familiar outside the realm of space travel—such as the rubber O ring, which you'll find on garden hoses and SCUBA gear, and which also happens to be used as a sealant between segments of rocket fuel boosters, and it's included in this book because it's responsible for perhaps the worst tragedy in the history of space exploration: the space shuttle *Challenger* disaster. At the other extreme of this spectrum is the Large Hadron Collider—said to be the most complex machine ever built, it changed the way we understand the creation of the cosmos.

After reading *Space Exploration*, one comes away with an enormous sense of the acceleration of human ingenuity: The interval between the invention of the first two landmark objects is more than thirty thousand years—while only a couple of years separate the last two. The message is clear: Humans can accomplish anything we set our minds (and resources) to. Abundant challenges may be ahead of us, but these 100 objects make one wonder: *Are there any limits to what we can do?*

JOHN MATHER
April 2019

JOHN MATHER won the 2006 Nobel Prize in Physics for measuring the Big Bang. He is the senior project scientist at the James Webb Space Telescope, which is the successor to the Hubble Space Telescope.

INTRODUCTION

The cosmos is nothing if not vast, and its history is long—our current best estimate of its age is just shy of fourteen billion years. Up against the universe's incomprehensible scale, our brief history of exploring and understanding space can seem, well, modest, even negligible. The overwhelming majority of what's out there remains entirely unknown to us.

But that hasn't stopped us from looking. Our discovery of the nature of our universe and its evolution through time is probably one of the most spectacular human stories that can be told. Archaeological evidence shows that for many tens of thousands of years—if not longer—human curiosity has driven us to dream up realms beyond our physical world and, just as important, to record our discoveries, which we're now uncovering through the artifacts that previous civilizations have left behind. Ancient lunar calendars, star clocks, crystal lenses, and other prehistoric objects may not be the first instruments that come to mind when one thinks of our history of space exploration, but without them, there simply wouldn't be a space history.

All of which is to say: This is no ordinary space book. The 100 objects in this book showcase not the greatest hits you're already familiar with but rather the workhorse tools and game-changing technologies that have altered the course of space history and yet, in many cases, haven't become household names.

To pick the 100 most notable objects in space history is, of course, an impossible task; not only because one could just as easily fill thousands of pages with remarkable objects worth knowing about, but because any kind of ranking of their relative importance is an inherently subjective endeavor. But I'm a scientist, and so I've selected the tools and devices that, taken together, represent the major scientific discoveries—and celebrate the human ingenuity—of space technology, showing the ways physics and

engineering have brought about our greatest leaps in understanding the way our universe works.

Everyone knows about Neil Armstrong's first steps on the Moon—without a space suit, he would have had to stay in the lunar lander. We've all seen the iconic *Earthrise* image: the perspective-shifting photo of our world from afar—without the Hasselblad camera, the photograph wouldn't have been taken.

And so on. These 100 objects have changed the face of space exploration, and it's entirely possible that many—if not most of them—are landmark objects you've never even heard of before. They make it clear that we have made giant strides in our quest to search ever more deeply into the farthest reaches of the universe—and behind each new discovery is an object that expands our appreciation of space as well as the boundless imagination and resourcefulness we carry within us.

STEN ODENWALD

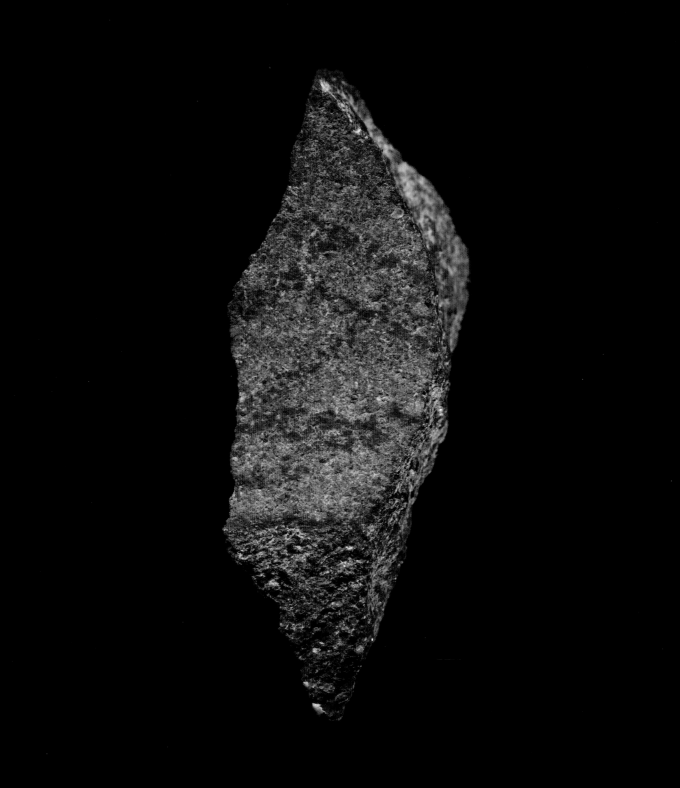

1

The Blombos Ochre Drawing

The first step to comprehending space

71,000 BCE

The first step on our journey toward a deeper understanding of the cosmos must begin long before we ever developed the ability to reach space. The immensity of the universe goes so far beyond our tangible world that to even begin to wrap our heads around it, we had to learn how to transform our surroundings into symbols and abstractions. And since what we'd come to learn about the cosmos would exceed any single human brain or life span, to make any real progress in our exploration of space we had to learn to build a permanent body of knowledge by recording what we've learned and passing it on to the next generation of explorers. We cannot know what our ancestors grasped before they invented a language large enough to encompass the many marvels of space, but at least we can find hints that our ancestors were traveling a path that would eventually lead them to a quantitative understanding of their world.

In 1991, in the Blombos Cave—located about 190 miles east of Cape Town, South Africa—archaeologist Christopher Henshilwood (now at the University of Bergen) and his team uncovered traces of Stone Age *Homo sapiens* inhabitants dating from as early as 100,000 BCE. The cave had been occupied multiple times, with each set of residents leaving behind shells, spear points, and some bone tools. But the most remarkable among these finds was discovered two decades later, when a research fellow who was cleaning the artifacts stumbled upon a small stone flake, about $1^{1}/_{2}$ inches long and $^{1}/_{2}$-inch wide, covered in distinct red lines. Henshilwood's team eventually determined that the lines were applied using a kind of crayon made of ochre, a natural pigment, roughly 73,000 years ago.

As to what the lines actually meant to the people who made them, it's impossible to say. But the crisscross markings seem intentional enough that many archaeologists interpret them as deliberate visual representation, making the stone the earliest known drawing done by a human hand.

Whatever the lines were meant to illustrate, the significance of this simple drawing cannot be denied. It gives us a firsthand look at the roots of our use of symbols, which would make possible both written language and math. And so, in some ways, it represents a big bang of human ingenuity, a starting point from which an explosion of knowledge would follow. Eventually, our abstractions turned toward the stars: Some experts believe that the spectacular renderings of animals at the famous cave in Lascaux, France—inhabited as far back as twenty thousand years ago—include figures and dot patterns that mark constellations along what we now call the zodiac: the band of star patterns through which the Sun and planets move each year. If true, by then our distant ancestors were keen observers of the sky.

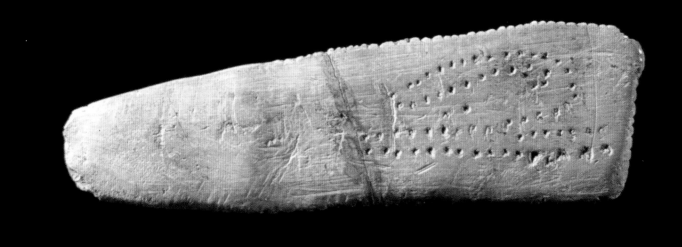

2

The Abri Blanchard Bone Plaque

Ancient lunar phase calendar

30,000 BCE

O ur prehistoric ancestors lived rather precarious lives. Thirty thousand years ago, at a time when hunter-gatherers couldn't be certain of their next meal, they likely spent a good deal of their time following the migrations of the animals that were their principle food sources. Animals follow patterns of movement over the course of the seasons and in cadence with systematic changes in the local weather and temperature. Edible plants and berries follow the rhythms of the growing seasons, too.

But what does all this have to do with space exploration? Arguably, it was the unpredictability of our ancestors' food supply that compelled them to develop rudimentary science, which is, in many ways, a tool for making predictions. Our ancestors undoubtedly probed the world around them in search of recurring patterns by which they could anticipate nature's cyclical course.

Perhaps the richest and most reliable set of patterns they would've found was overhead: The Moon's shape seemed to change over the course of twenty-nine (or so) days before starting all over again. The Sun rose in one direction (east) and set in the opposite direction (west) and never the other way around. There were stars in the sky whose constellations slid constantly westward from month to month, but the constellations themselves remained fixed. What today we call Orion always looked like Orion; Scorpius always looked like Scorpius. And the entire sky seemed to rotate around a fixed point in the sky every night, dependably revealing direction in the North Star and serving as a beacon in the winter for travelers looking for warmer climates, and for cooler climates in the summer.

We can't know exactly what meanings our ancestors read into the movements of celestial objects, but we can be far more confident that they took careful note as far back as 30,000 BCE, because we have strong archaeological evidence—numerous renditions of lunar shapes and counting systems of the twenty-nine-day lunar cycle have been found from this era on antlers and other media. Perhaps the most striking artifact is the Abri Blanchard bone plaque, named after the cave in southwest France where it was found. Carved into a flat bone fragment is a sequence of notches that gradually wane and wax between crescent shapes and full circles. Some experts have even read further into the sequence, suggesting that the markings move in sets of seven from crescent to half circle, from half to full, and then from full to half and then to crescent again. This is only a theory, but in any case, this bone plaque makes a persuasive case that our ancestors thought it important to create a permanent record of a natural, predictable cycle—the kind of thinking that sets the stage for the pageant of scientific discovery and advancement to come.

▲
A star clock on a wooden coffin lid from the Eleventh Dynasty in Asyut, Egypt

3

The Egyptian Star Clock

The first steps toward quantifying the sky

2100 BCE

The ancient Egyptians were skilled timekeepers, and they left behind a wealth of objects that could each in themselves be considered major milestones in how we came to understand the movements of stars and our Sun. Obelisks, some more than four thousand years old, show the march of time by shadow, and it's not much of a technological leap from there to the first known sundial, from the thirteenth century BCE, found in the Valley of the Kings near Luxor.

But it would be almost unfair to make too much of the sundial when by as early as around 2100 BCE, ancient Egyptians had developed a far more technologically impressive timekeeping system in the decan—a sequential series of thirty-six constellations that would be used for timing the hours in a

day and the days in a year. A new decan would be visible just before sunrise every ten days, but five additional festival days were added to create the full calendar year. The year began with the appearance of the first decan, the rising of the constellation Sirius (*Sopdet* to the Egyptians), which heralded the all-important, life-giving flooding of the Nile. At night, in accordance with Earth's rotation, a new decan would rise every forty minutes, defining a decanal hour. This decanal calendar system is probably older than 2100 BCE, but this is roughly the age of the oldest surviving recorded appearance.

Decanal star clocks began appearing on coffin lids during the tenth pharaonic dynasty (ca. 2160 to ca. 2040 BCE). Ancient Egyptians did not concern themselves with detailed depictions of the shapes of the decan constellations, instead often representing them by a simple list of star hieroglyphs in thirty-six columns, ostensibly one for each bright star in the decan. The image above shows the coffin lid from the Eleventh Dynasty tomb of Idy, a distinguished official, from Asyut, a city along the Nile in central

Egypt. Spanning nearly the entire length of its central panel are the decans, lined up in neat rows.

The star clock is yet another example of the sustained interest in celestial objects by ancient peoples, and it represents a giant leap in our capability to track and predict their cycle. Furthermore, the coffin decans are the first documented attempt at quantifying what we can see in the sky—the basis, in the coming millennia, for modern astronomy and astrophysics.

The first known sundial, from roughly the thirteenth century BCE, found in the Valley of the Kings in Luxor, Egypt ▶

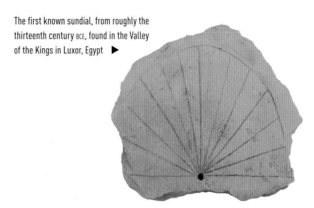

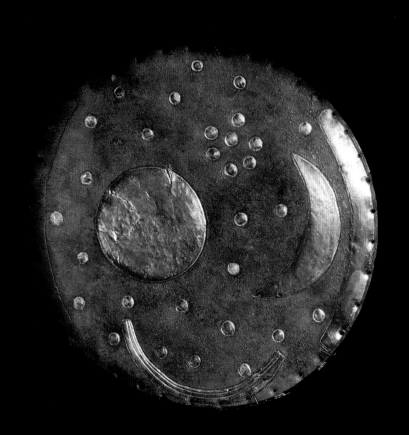

4

The Nebra
Sky Disk

A pocket planetarium

1600 BCE

The Nebra sky disk, a bronze medallion about twelve inches in diameter and weighing just under five pounds, is so singular in its design that it was initially considered a forgery. Discovered by two amateur treasure hunters in 1999 in a forest in central Germany, it was illegally removed and sold to a dealer in Cologne. Through various police operations, it was retrieved in 2002 by a state archaeologist and now resides in Germany's State Museum of Prehistory in Halle.

A careful study of the patina of green rust revealed the disk was not a forgery at all but rather an artifact of considerable age. Radiocarbon dating of a piece of birchbark found on a nearby item at the excavation site suggested a burial age between 1600 and 1560 BCE, though technically the Nebra Disk could have been manufactured decades or even centuries before it was buried. It's a stunning work of art, of course, but it earns its place as a milestone in space history for a number of key details that appear to go beyond Bronze Age artistry.

The disk was meticulously crafted to depict careful—and remarkably accurate—celestial observation. First, a circular plate was fashioned to represent the full Moon or Sun, a crescent-shaped Moon, and dots for stars, with the seven Pleiades stars added in a cluster—the easiest star cluster to see with the naked eye. Secondly, two arcs frame the disk on opposite sides, each spanning eighty-two degrees. This angle roughly matches the angle between the sunsets at the summer and winter solstices at the latitude of the excavation site.

It's widely known that enormous monuments such as Stonehenge in England seem to have been constructed in careful celestial alignment, and that therefore our ancestors had, by several thousand years ago, recorded with precision the movements of the Sun and Moon. But the Nebra is the earliest example of a portable device for tracking the solstices, suggesting that, by the Bronze Age, awareness of the movements in the heavens was a necessary part of daily life—perhaps to help manage crop production.

The disk is also a landmark in space history because of the way it looks. It's the earliest example we have of the Sun, Moon, stars, and sky depicted realistically.

5

The Venus Tablet of Ammisaduqa

Modern astronomy's foundational text

1500 BCE

The Venus Tablet of Ammisaduqa, tablet 63 of the Babylonian series on celestial observations, Enuma Anu Enlil, records in cuneiform (one of the world's first writing systems) the rising times of Venus as well as its first and last visibility on the horizon at sunrise or sunset for a period of about twenty-one years—in Year 1, for example, it says that "Venus sets on Shabatu [the eleventh month of the Babylonian calendar] 15 and after 3 days rises on Shabatu 18." It resides in the British Museum and measures roughly seven inches tall, four inches wide, and about an inch thick. It is one of many cuneiform tablets that were a part of the library of King Ashurbanipal and one of over thirty thousand items excavated from Nineveh, Iraq, in the 1850s.

Ammisaduqa, of the First Dynasty of Babylon, was the fourth king to rule Babylon after Hammurabi, and he enjoyed a peaceful twenty-one-year reign. The planet Venus played an important role in Babylonian mythology, being associated with Ishtar, the goddess of love, sex, war, fertility, and political power. Predicting the comings and goings of the planet was critical to divining omens on the king's behalf, lending utmost importance to keeping careful observations and records.

The Venus Tablet is an extraordinary foundational text of astronomy: With its predictions of Venus's rising and setting times across more than two decades, it's the first known evidence that humans have of an understanding that celestial phenomena can recur at regular intervals. To make such predictions required the use of math, too—also a first. Without either breakthrough, modern astronomy simply wouldn't exist.

6

The Star Charts of Senenmut

Drawing the sky in detail

1483 BCE

In some ways, while our methods for seeing and understanding space are dramatically sharper, we live in something like a dark age of consciousness about the cosmos: Many of us don't spend much time looking up at the stars. Or perhaps "light age" is a more appropriate name for our era—with growing urbanization and the wide spread of artificial lighting, there are very few places on Earth that aren't significantly affected by light pollution. The result is a sky largely devoid of detail; for those of us who stargaze, there's not nearly as much to see as there once was.

In this way, the ancient star charts of Egyptian architect and chancellor Senenmut represent something of a peak in celestial influence on daily life. These detailed depictions of the skies reveal how, to ancient Egyptians, the planets and stars were everything.

Senenmut entered into royal service during the reign of pharaoh Thutmose II and continued as the high steward during the reign of the queen-pharaoh Hatshepsut. He is thought to have been the architect of Hatshepsut's magnificent mortuary temple at Deir el-Bahri, ca. 1479 to ca. 1458 BCE. The ceiling of his unfinished tomb shows numerous elaborate drawings of the sky and are believed to essentially summarize everything Egyptians knew about the sky and their calendar system by the Eighteenth Dynasty. The most impressive renderings are on the two ceiling panels, which show constellations in detail. The three nearly vertical stars at the center of the top panel represent the belt stars in Orion (called *Osiris* by the Egyptians), as well as Osiris, who is

personified in a boat just below the constellation. His sister and wife, Isis (Sirius), in the column immediately to the left, stands with a crown of two feathers. Between Isis and the column containing the turtles are the two figures of Horus, the son of Osiris and Isis, representing the planets Jupiter and Saturn. In the far-left column is a Bennu bird carrying Venus.

Separating the top and bottom panels are a five-line prayer to Senenmut. Below this is the second panel, with its twelve circles. It's one of the most photographed drawings of this period due to its distinct geometric forms. The circles correspond to the twelve months of the lunar year. Each is divided into twenty-four sectors, presumably one for each day. The central, vertically aligned artwork depicts the circumpolar constellations, starting with the top figure, the Bull (Ursa Major)—you can make out the stars in its tail. The Bull is looking at the falcon-headed god Anu holding a spear, probably representing Cygnus, the Swan. At the bottom is a more complicated drawing of a man fighting a crocodile, identified with some stars in the constellations of Draco and Ursa Minor. The hippopotamus figure, Isis-Dyamut, is shown to the right with a crocodile on her back, believed to be stars in the constellations Boötes, Lyra, Hercules, and Draco.

These drawings are something like a Rosetta stone for Egyptian astronomy; without them, we would scarcely know how they viewed their astronomical universe, how one realm—the stars and planets—related to the others, enmeshing their sense of time with their gods, who, they believed, held sway over every part of daily life.

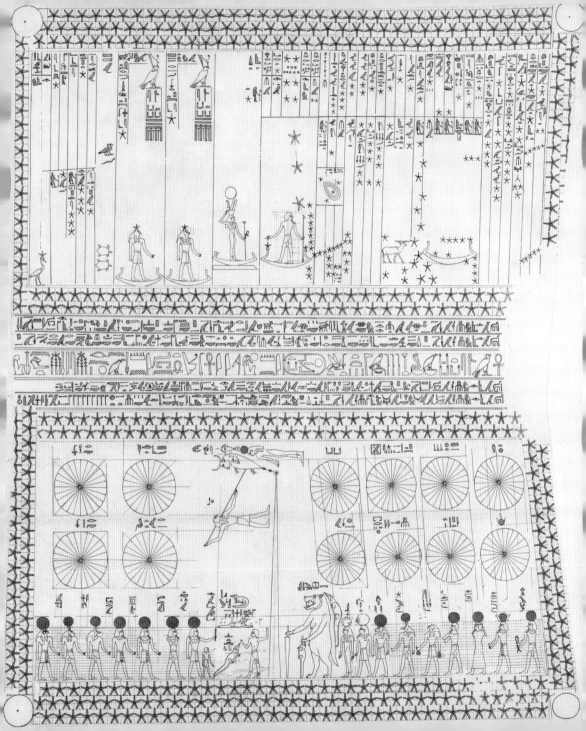

7

The Merkhet

**Fusing astronomy
and construction**

1400 BCE

The farther back in time we look, the less we know about our ancestors; the passage of time wears down and destroys many of the objects they left behind. One of the primary kinds of artifacts left is a civilization's monuments, which tend to be built to last. And from ancient Egypt's monuments, we can infer that they used some type of measuring instrument to make sure lines were straight and angles were true. For example, the triangular faces of their pyramids are pitched at a fifty-two-degree angle, known as the *seked*, the precision of which suggests that they used some triangular instrument or framework during construction. But these instruments, typically made of wood and string, were perishable, and so are now largely lost.

Fortunately, some evidence remains, preserved and recovered in tombs. Among the most basic items found are the square, the plumb, and

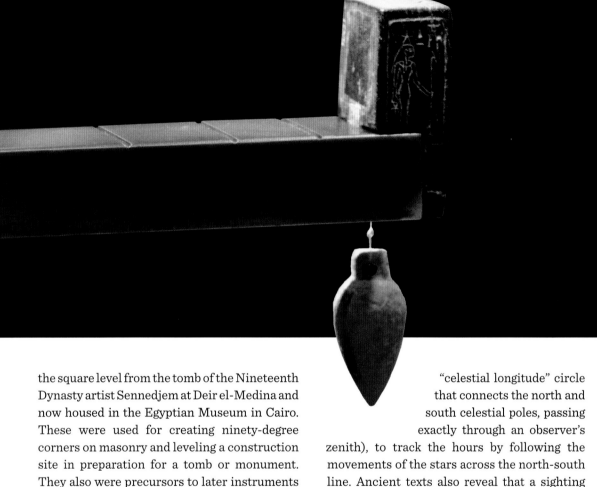

the square level from the tomb of the Nineteenth Dynasty artist Sennedjem at Deir el-Medina and now housed in the Egyptian Museum in Cairo. These were used for creating ninety-degree corners on masonry and leveling a construction site in preparation for a tomb or monument. They also were precursors to later instruments used to measure star locations in the sky with increasing precision.

One of the most striking among these tools is the *merkhet*, whose name means "instrument of knowing." It consists of a bar and weighted tether that could hang to the ground to determine the axis and astronomical alignments of buildings.

But more important in the context of space, it was also used to gauge the passing of time at night. Two merkhets would be used simultaneously, one aligned with the North Star, the other with a north-south meridian (the "celestial longitude" circle that connects the north and south celestial poles, passing exactly through an observer's zenith), to track the hours by following the movements of the stars across the north-south line. Ancient texts also reveal that a sighting tool was used along with the merkhet to find north and map the skies. It was a breakthrough in the way we understand space because of the leap forward in accuracy it allowed regarding astronomical measurement.

The merkhet pictured here, housed in the Louvre, depicts Amenhotep III offering Maat, the goddess of order and harmony, to the Sun god. It's dated as far back as 1400 BCE.

8

The Nimrud Lens

The first step toward modern telescopes

750 BCE

Telescopes, one of the most important inventions in all of astronomy, may call to mind the complex instruments in use today. But the basic elements behind a refracting telescope's operation haven't changed for millennia.

That includes the central component, the lens. The earliest-known lenses were made from polished crystal, often quartz. And the most ancient lens in the world is the Nimrud lens, dated to 750 to 710 BCE and discovered in the ancient Assyrian city of Nimrud, in present-day Iraq, in 1850, by the English archaeologist Austen Henry Layard. It is a disk of crystal about 0.5 inches in diameter and about 0.1-inch thick. It has been polished so that it has a rough focal length of 4.72 inches. The optics make it equivalent to a 3x magnifying lens. It isn't entirely known how this lens was used, but it was clearly fabricated and could have been used to concentrate sunlight for starting fire or, as some archaeologists suggest, as nothing more than a decorative amulet of some kind. However, very small inscriptions were found on other nearby artifacts, and these, the theory goes, might have been made using this lens as a magnifier.

It's not much of a magnifier, but if the prevailing theory is true, it's a landmark in optics. A lens works by refracting light, bending rays through its curved surface so that the source of the light appears farther or, in the case of a convex lens like the Nimrud, closer than it actually is. It's the same property used by refractive telescopes: Light from space is refracted through a lens so that distant objects appear in far greater detail, as if before our eyes.

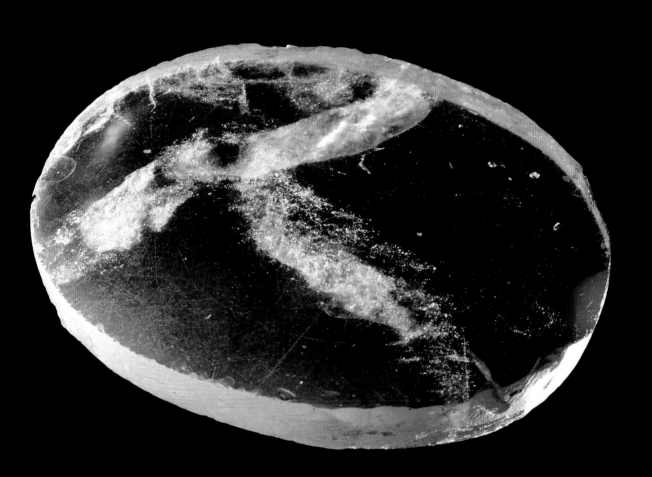

9

The Greek Armillary Sphere

The first celestial calculators

300 BCE

An armillary is a rotatable, spherical shell made of rings usually tilted on an axis matching Earth's 23.5-degree polar inclination. It is crisscrossed by circumferential rings, or bands, representing the celestial equator (Earth's equator projected into space), the ecliptic (the plane of the path the Sun, Moon, and planets follow), and the north-south meridian. Additional circles for the Tropics of Capricorn and Cancer and for the Arctic and Antarctic Circles have also been added to later models. Within this shell is usually a small globe representing Earth.

The earliest armillary spheres are credited to the ancient Greeks; the astronomer Hipparchus identified Eratosthenes (ca. 276 to ca. 194 BCE) as the inventor. It was independently invented in China by the astronomer Zhang Heng, who lived from 78 to 139 CE. So, by the third century CE, it had become a tool of astronomers from west to east, who would use them to perform calculations related to the movements of the Sun, Moon, and planets across the sky. They were usually aligned with the axis of Earth so that the sphere could be rotated to show the locations of the various bands in the local sky. Some were even automated with a clockwork mechanism so that they rotated in step with the turning of the sky every day. Over time, they became a popular teaching tool, and artists often incorporated them into portraits of their benefactors to imply intellectual sophistication.

Although we know they were in wide use since the second century BCE, these mechanical instruments have a poor record of surviving intact through the Middle Ages and into the Renaissance. One of the oldest surviving armillary spheres is the extremely elaborate one constructed by Antonio Santucci in 1582, located in the library at the El Escorial monastery near Madrid. Even older paintings of armillary spheres still survive, such as the one by Justus of Ghent in 1476, which shows Ptolemy holding one.

◀ Antonio Santucci's armillary sphere

10

The Diopter

A landmark in charting accurate star positions

200 BCE

One the most game-changing ideas ever to emerge in astronomy was the notion that the positions of the stars could, and should, be measured. Mapmaking by surveying the land is an ancient skill dating to over ten thousand years ago. All that was required to map the heavens was that ancient surveyors literally turn their equipment skyward. We've seen the earliest example of this in the Egyptian merkhet; the Greeks would take this a step further and start to make measurements. The first indication of this dates to the ancient Greek astronomers who used an instrument called a *dioptra* to measure the positions of stars. Both Euclid, ca. 300 BCE, and Geminus, ca. 70 BCE, refer to this instrument

◀ A reconstruction of Heron's dioptra

in their astronomical works. Unfortunately, today, all that remains in surviving manuscripts are mentions of what these instruments looked like. Heron of Alexandria (ca. 10 to ca. 70 CE) wrote an entire book on how to build and use a diopter for surveying. Many attempts to create the instrument using his method have been made, and it is clear that it could have been used to create accurate star atlases and maps.

Heron's dioptra consisted of a tripod upon which a circular, rotating table was mounted. Through a combination of adjustable screws, water levels, and a sighting tube attached to the table, one could spot a celestial object, then turn the table and spot another, and determine the precise angle between the two. The device could also be used to find the elevation angle of a star above the horizon.

The diopter eventually gave way to the theodolite, first mentioned in the 1571 surveying textbook *A Geometric Practice Named Pantometria*. After the invention of the telescope, the original sighting tube was replaced by a small telescope, which was attached to a two-axis moving framework. The vertical axis moved on a graduated semicircle divided into degrees with a caliper that could make measurements to fractions of a degree.

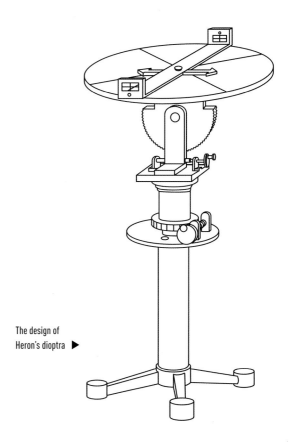

The design of Heron's dioptra ▶

A theodolite built in 1851 ▶

But the basic technology is the same: something to look at the skies with, affixed to something that can measure the angular distances between what you see.

The diopter, and the instruments that followed it, were crucial in setting the stage for developing accurate star maps, and they remained in use until twentieth-century technologies replaced them. They represent a landmark in the way we mapped the sky: Gone forever were eyeball estimations, replaced by careful measurement, the cornerstone of virtually every space-related technology and discovery to come.

11

Antikythera Mechanism

A portable celestial calculator

200 BCE

In 1901, a curious artifact was recovered from an ancient shipwreck off the coast of the Greek island of Antikythera. A year later archaeologist Valerios Stais found that it contained a gearwheel. The gear was part of a single bronze item mixed with wood from the remains of a box, and it measured only 13 × 7 × 3.5 inches. Stais initially proposed that it was a seafaring instrument, but most scholars thought it was too advanced for the estimated date of the shipwreck, from roughly 60 to 205 BCE. No further studies on the instrument were undertaken until the second half of the twentieth century, when science-historian Derek J. de Solla Price and physicist Charalampos Karakalos used X-ray imaging techniques to detect as many as thirty-seven gears.

From the gear ratios and what remained of the framework, Price and Karakalos deduced that it was very likely an analog computer that could be used to predict the motions of the Sun and Moon, the lunar phases, eclipses, and the dates for the Olympic Games, among other events. By setting the correct solar date on the front panel, the rear panel would display the correct lunar month to an accuracy of a week or so. And while there are no gears or combinations that would correspond to any of the planetary movement cycles, there are pointers for the five planets known at that time, so it may be that the gear works for the planetary calculator are missing.

Since 2005 the members of the international Antikythera Mechanism Research Project have been working to learn more about its structure, applications, location of manufacture, and when and by whom it was constructed. But we know enough now to appreciate the astounding device as one of the greatest, most momentous inventions of all time, and not only for its remarkably complex predictive capabilities. It's recognized as the world's first analog computer—so modern life would be unrecognizable without its groundbreaking technology.

Reconstruction of the complete mechanism ▶

12

Hipparchus's Star Atlas

A foundational map of the heavens

129 BCE

An engraving of
the Farnese Atlas ▶

Hipparchus is one of the ancient world's most famous astronomers. He's the one who discovered the phenomenon of precession, the way Earth's axis gradually shifts over a period of about twenty-six thousand years, which shifts the alignment of the stars as we see them in the sky. But while he wrote at least fourteen widely cited books on astronomy and mathematics, his only surviving work is his *Commentary on the Phaenomena of Eudoxus and Aratus.* Everything else, and in particular his atlas of the sky, is lost.

The Hipparchus star catalog is believed to have contained 850 bright stars. It was completed sometime late in his life, perhaps around 129 BCE. His astronomical research was later incorporated into the star catalog published in Ptolemy's book *Almagest,* the supremely influential work of astronomy that established the geocentric model of the universe, sometime around 150 CE.

Ptolemy notes in his catalog of 1,020 stars that the longitudes had increased by two degrees and forty minutes since the time of Hipparchus, which is the expected precession of Earth's rotation axis since the time of Hipparchus. This strongly suggests that it

was largely the Hipparchus star catalog that Ptolemy used, but he added two degrees and forty minutes to the longitudes. Even more reason, then, to consider Hipparchus's star catalog a seminal lost work: It served as the basis for Ptolemy's own atlas.

Then, miraculously, in 2005 astronomer Bradley Schaefer at Louisiana State University announced that the Hipparchus star catalog had been hiding in plain sight—or at least a representation of it. A second-century Roman statue called the *Farnese Atlas* at the National Archaeological Museum in Naples, Italy, shows Atlas carrying a globe of the heavens on his shoulder. The globe depicts, in relief, forty-one constellations placed on a grid of circles representing the equator and the Tropics of Capricorn and Cancer. Through a careful study of the locations of the constellations, Schaefer was able to show that to accurately draw them, the sculptor would have had to use a star catalog and constellation guide from around 125 BCE—approximately the time when Hipparchus would have created his catalog. So there it was, an artistic rendering of Hipparchus's long-lost work, held up for nearly two millennia by the arms of Atlas.

13

The Astrolabe

Using the stars to track time

375 CE

Astrolabes were the smartphones of the ancient world—all-in-one contraptions that could tell you the time and your location, among other things. They're measuring devices consisting of a circular star map that rotates so it displays the stars visible at the specific latitude for which the astrolabe was designed. They also came equipped with movable circles and pointers to indicate the brightest stars and the ecliptic. A spotting tube was added so that the astrolabe could be used as a diopter to measure the elevation of stars above the horizon.

The time line for the development of the astrolabe is murky—it puts into use mathematical principles developed over centuries—but the first source to directly discuss the astrolabe is Theon of Alexandria's "On the Little Astrolabe," dating to sometime around 375 CE, which set the standard for subsequent writings on the tool. However, the astrolabe came to be, it caught on primarily as a means to measure the elevation of stars above the horizon, and to determine an observer's latitude

by sighting the polestar. Astrolabes were introduced to the Islamic world by 800 CE, where they were dramatically improved upon by adding angular scales as well as circles to determine azimuth, making them invaluable for navigation and, more important, keeping track of the direction to Mecca.

▲
An astrolabe
from ca. 1400

Astrolabes were in many ways equivalent to the modern slide rule or calculator, and they imbued the user with a kind of mystique. After all, they could tell you your latitude from all over the northern hemisphere by locating the North Star—an incredible power in a single tool. And if you knew your latitude beforehand, you could use it to tell local time by sighting a few key stars and consulting the handy graph or gridwork inscribed on the plate. The instruments were even used by Ptolemy to make astronomical sightings for his famous book *Tetrabiblos*, his writings on astrology. Over the centuries, many authors penned detailed treatises on how to build and use astrolabes and on their operating principles. These dynamic devices are marvels in themselves, and they were also yet another major stepping-stone in the progression toward ever-more-precise astronomical predictions.

31 ///

自斗二度相風痙其七度於辰莊丑為星紀者言統己万物之終故曰
歳之分也

天横

天柱

文昌

天掊

立

相

大陽首

The Dunhuang
Star Atlas

The first complete star chart

700 CE

The Mogao Caves, or Caves of the Thousand Buddhas, are a system of 492 caves near the city of Dunhuang in the Gansu Province of China. Between the fourth and fourteenth centuries, a complex network of caves was dug out along the Silk Road by Buddhist monks to serve as shrines. After the end of the Yuan Dynasty in 1368 CE, the area went into decline and was abandoned until the late 1800s, when archaeological interest in this Silk Road site began to increase. Chinese Taoist priest Wang Yuanlu began excavating and restoring the site, and on June 25, 1900, he unearthed a small cave filled with thousands of manuscripts. A lack of interest in the find by the Chinese government led to thousands of these documents being removed by foreign archaeologists and dispersed to archives in London and elsewhere. One document was a parchment scroll acquired by the Hungarian British archaeologist Aurel Stein in 1907. Ten inches wide and 155 inches long, it now resides in the British Library.

It would be several decades before anyone understood the parchment's significance—its first mention in the astronomical literature didn't come until 1959, in Joseph Needham's book *Science and Civilisation in China*. And although Chinese historians and astronomers have studied the manuscripts since the 1960s, they did not have access to the original scroll and therefore worked with published photographs. It wasn't until 2009 that the star chart was analyzed in considerable detail by the French astrophysicist Jean-Marc Bonnet-Bidaud.

The Dunhuang Star Atlas is generally recognized as the first-known complete star atlas still in existence, dating from before the Tang Dynasty, circa 618 CE. We know that there were other star atlases and catalogs in the ancient world, but none besides this have survived to the current era. This atlas, likely created by astronomer Li Chunfeng, shows the positions of 1,339 stars grouped into 257 constellations and asterisms (popular groupings of stars that are smaller than constellations) based on data provided by three ancient astronomers: Wu Xian, Gan De, and Shi Shen, whose contributions in the chart are color coded. Consisting of twelve star maps in total, the atlas's bright stars are rendered in position to an accuracy within a few degrees. The thirteenth map, shown here, covers the northern circumpolar constellations, while Map 6 names the bright star Laoren (the Chinese name for Canopus), a southern star that suggests Chinese astronomers had investigated the southern sky. Map 5 contains the easily recognizable asterisms related to Shen (Orion).

With astonishing accuracy, the atlas is consistent with present-day maps. The stars were not randomly placed for artistic effect but rather followed a mathematically outlined plan. Many of the previous objects in this book helped to bring about this achievement: the skies charted accurately and comprehensively.

15

Al-Khwārizmī's Algebra Textbook

Multiplying the power to calculate our universe

820 CE

A page explaining how to use algebraic geometry to calculate the area of quadrilaterals, including the square and rectangle ▶

One way to think of algebra is as a bridge between the abstract, representational iconography we've largely seen up until this point and the razor-sharp mathematics that lie ahead. The word *algebra* itself comes from the Arabic *al-jabr*, which is translated as "the reunion of broken parts." This comes from the title of the book *Ilm al-jabr wa'l-mukabala*, written by the Persian mathematician and astronomer al-Khwārizmī ca. 820 CE. Although al-Khwārizmī did not develop the entire field, he at last brought many of the ancient ingredients together in one tome. Although a key ingredient in algebra is the replacement of numbers with letters, especially x to represent an unknown quantity, this usage did not become widespread until the time of Descartes. In his book *Les Géométrie*, published in 1637, he used letters like a, b, and c to represent known quantities and letters at the other end of the alphabet for unknown quantities, including x—the first documented case of this practice.

At its core, algebra is a system of symbols that stand in for unknown quantities but nevertheless obey the basic operations of addition, subtraction, multiplication, and division. Algebra's greatest strength isn't its usefulness in determining specific answers to individual problems but rather the way it can be used as shorthand to describe a procedure, called an *algorithm*, for finding the answer to a whole *type* of problem, regardless of what the actual numbers are in a given situation.

To put this in concrete terms related to space: The universe is far from static—for instance, every star, planet, meteor, moon, and more are in constant motion in relation to one another. With so many ever-changing variables, it would be a slow and inefficient slog to make calculations based on single data sets, only to have to do them again and again with each change in data. Algebra is the key to unlocking the potential of engineering and physics because it allows us to calculate movement and forces in their natural, dynamic, ever-changing states— paving the way for the breathtaking pace of technological advancement we've grown so accustomed to today.

◀ Pages from al-Khwārizmī's algebra book

أعلم أن المربعات (١) خمسة اجناس فمنها مستوية الاضلاع قائمة الزوايا والثانية
قائمة الزوايا مختلفة الأضلاع طولها ا كثر من عرضها . والثالثة تسمى المعينة وهى
التى استوت اضلاعها واختلفت زواياها . والرابعة المشبهة بالمعينة وهى التى طولها
وعرضها مختلفـان وزواياها مختلفة غـير أن الطولين متساويان والعرضين
متساويان أيضاً . والخامسة المختلفة الاضلاع والزوايا . فا كان من المربعات
مستوية الاضلاع قائمة الزوايا أو مختلفة الاضلاع قائمة الزوايا فان تكسيرها
أن تضرب الطول فى العرض فما

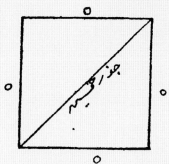

بلغ فهو التكسير . ومثال ذلك
أرض مربعة من كل جانب خمسة
أذرع تكسيرها خمسة وعشرون
ذراعاً وهذه صورتها . والثانية
أرض مربعة طولها ثمانية أذرع

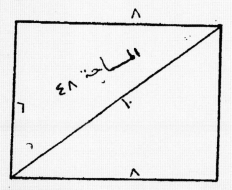

ثمانية أذرع والعرضـان ستة
ستة . فتكسيرها أن تضرب
ستة فى ثمانية فيكون ثمانية
وأربعــين ذراعاً وذلك
تكسيرها وهذه صورتها .
وأما المعينـة المستوية
الأضلاع التى كل جانب منها

---

(١) أى الاشكال الرباعية بالاصطلاح الحديث وتقسم هنا إلى مربع ومستطيل
ومعين ومتوازى أضلاع وشكل رباعى عام .

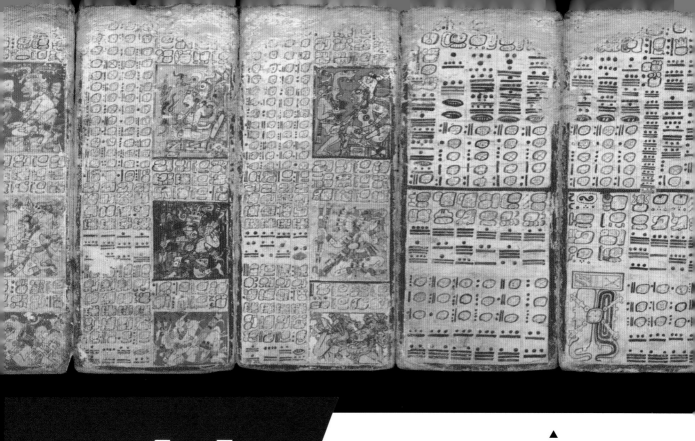

16

The
Dresden Codex

A glimpse at Mayan
precision astronomy

1200 to 1300

O ur knowledge of astronomical science in the European "Old World" is vast and deep thanks to a surfeit of monuments, writings, books, and inscriptions scattered across the continents. By comparison, our knowledge of the various civilizations that have come and gone across the Americas is quite limited. In regions where human activity was the most intense, dense and inaccessible jungles now block the way. But more important, whatever written documents were present among the many Mayan and Incan civilizations were virtually all destroyed by the conquistadors in the sixteenth century and the aggressive missionary movement that followed.

The *Dresden Codex*, then, is an outlier, the only surviving written record from the Mayan empire

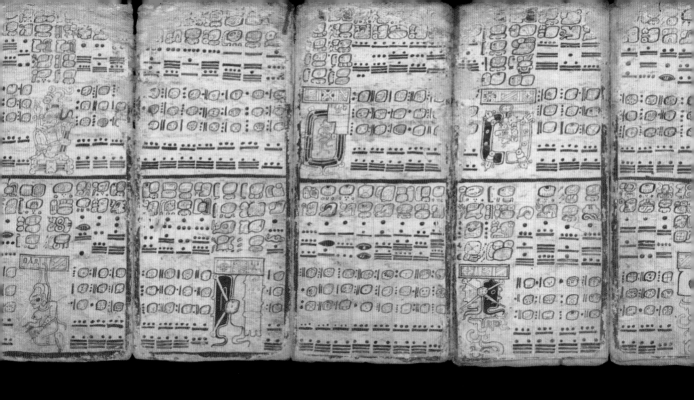

ca. the fourteenth century CE. Johann Christian Götze, director of the Royal Library in Dresden, Germany, purchased it from a private owner in Vienna, Austria, in 1739. Archaeological studies of unique symbols in the seventy-eight-page codex suggest it may have been written near Chichen Itza on the Yucatán Peninsula. There is also evidence that the residents there were conversant in sophisticated astronomical information by 1200.

The codex includes astronomical tables for Venus and the Moon, including information about lunar and solar eclipses. The Venus tables describe the movements of Venus through 65 synodic periods for this planet (584 days for each), of which the Mayans were keen observers. Venus was associated with the god Kulkulkan, and its apparitions in the sky were used to plan wars. In addition to tables of ritual schedules and astrological information, the codex includes a ritual cycle of 260 days called a *tzolk'in*, which is a non-astronomical cycle formed from the product of 13 months of 20 days each. For the Maya, their religious celebration days

were crucial, but since the solar year is 365.25 days, which they called the *haab*, they had to correct for this accumulating difference over time just as we add an extra day to February every four years. To make this correction, they used the movements and sightings of Venus.

Space is above us no matter where we are on Earth, of course. This codex is a one-of-a-kind window into how we understood astronomy up to this point in the western half of the world.

◄ The Chaco Canyon Sun Dagger
in a winter solstice alignment

The Chaco Canyon Sun Dagger

An homage to celestial movement in spiral and light

1300

There are numerous locations in the American southwestern deserts where ancient petroglyphs (rock carvings) can be found on remote stones and cave walls. Perhaps none of them is more iconic than the sun dagger found on Fajada Butte in Chaco Canyon, New Mexico. It was discovered in 1977 by a local artist, Anna Sofaer, while exploring the region. Tucked away under an ancient fall of rocks was a curious spiral-shaped petroglyph, fortuitously illuminated by a sliver of sunlight angled between the sides of two large rocks partially blocking the wall. This singular shaft of light soon became known as the Chaco Canyon Sun Dagger, and it appears only at the time of the summer solstice. At the winter solstice, however, two daggers of light bracket the spiral petroglyph. Light shafts also strike the center of a smaller spiral petroglyph located nearby. In 1989, however, it was discovered that the sandstone slabs had shifted in position, and so the Sun Dagger, seemingly in operation since about 950 CE, when the Anasazi occupied this area, no longer appears. Only the spiral petroglyph is left to mark this remarkable show. Had Anna Sofaer not stumbled upon it in 1977, we would never have known

◄ The dagger at
summer solstice

of its existence and function, thinking it just another petroglyph in an obscure and perplexing location.

Meanwhile, similar dagger displays marking the solstices and/or equinoxes can be found at other locations in the southwestern United States and in Mexico, such as Hovenweep National Monument in Colorado and Utah, Burro Flats in southern California, and La Rumorosa in Baja California.

The relative obscurity of the Chaco Canyon Sun Dagger reminds us that curiosity about the sky and the timing of the seasons is far harder to document among civilizations in which oral histories emerged rather than a written language. The indigenous North American tribes left evidence of sophisticated astronomical knowledge and its practical applications in agriculture only in the rare monuments and petroglyphs we find scattered among mostly little-traveled regions of the Great Plains and the southwestern deserts.

18

Giovanni de' Dondi's Astrarium

An astonishingly complex calculator from the late Middle Ages

1364

In 1364, the physician and amateur astronomer Giovanni de' Dondi completed sixteen years of work on a technological masterpiece: a clock that also displayed the movements of the planets. It was an intricate and complex mechanism consisting of 107 gear wheels and pinions in a brass framework. In many ways it resembled the Antikythera mechanism created by ancient Greek craftsmen more than 1,400 years earlier. The astrarium (planetarium) was regarded with such amazement in the fourteenth century that it was considered the Eighth Wonder of the World.

Although de' Dondi's original creation didn't survive the ravages of time, his detailed plans did, and replicas made to his exact specifications and calculations can be created by anyone willing to invest the time. Many attempts to achieve this level of craftsmanship were made over the ensuing centuries, but few of the facsimiles worked because of slight imperfections in machining the gears.

One of the first working modern replicas was made in 1961 to 1963 by the Milanese watchmaker

A reconstruction of
the astrarium

Luigi Pippa and presented in 1985 to the International Watchmaking Museum in La Chaux-de-Fonds, Switzerland. Working copies are also to be found at the Paris Observatory, the Science Museum in London, and the Smithsonian Institution in Washington, DC, among others.

As an astronomical clock, it summarized in a physical way the accumulated knowledge about the periodic movements of the planets. For example, for each day, the astrarium gave the times of dawn and sunset (at the latitude of Padua). It also predicted the "Sunday letter," the lettering systems used to determine the day of the week on certain dates, the dates of saints' days, and the dates of the fixed feasts of the Catholic Church.

For de' Dondi's astrarium, the pure math required to predict the apparent movements of the Sun, Moon, and major planets was distilled into one remarkable device that the nonscientist could easily understand.

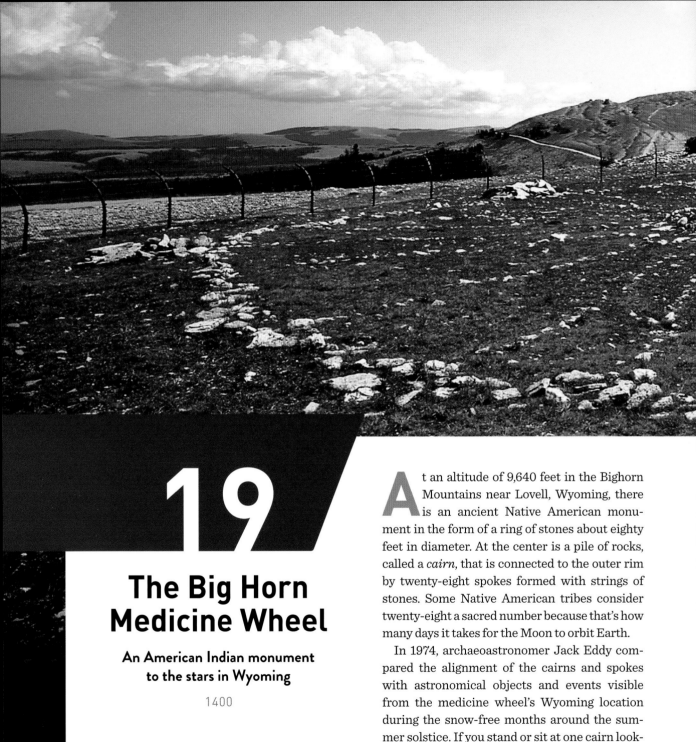

19

The Big Horn Medicine Wheel

An American Indian monument to the stars in Wyoming

1400

At an altitude of 9,640 feet in the Bighorn Mountains near Lovell, Wyoming, there is an ancient Native American monument in the form of a ring of stones about eighty feet in diameter. At the center is a pile of rocks, called a *cairn*, that is connected to the outer rim by twenty-eight spokes formed with strings of stones. Some Native American tribes consider twenty-eight a sacred number because that's how many days it takes for the Moon to orbit Earth.

In 1974, archaeoastronomer Jack Eddy compared the alignment of the cairns and spokes with astronomical objects and events visible from the medicine wheel's Wyoming location during the snow-free months around the summer solstice. If you stand or sit at one cairn looking toward another, your gaze will be pointed to certain places on the distant horizon. These points indicate where the Sun rises or sets on the

summer solstice and where certain notable stars—Aldebaran, Rigel, Sirius, and, as was later discovered by astronomer Jack Robinson, Fomalhaut—rise heliacally; that is, first rise at dawn after being behind the Sun. It seems, then, that these stars marked the summer solstice for visiting Native Americans.

The medicine wheel is located in the Crow homeland, in an area that their oral history says was given to them by an ancestor Crow leader whom historians have dated to a period between 1400 and 1600. This overlaps the precession age of the alignments, especially the one for the star Aldebaran between 1050 and 1450, giving us a rough construction period of about 1400.

The Big Horn Medicine Wheel is a stunning testament to the many ways humans have been inspired to record and predict the movement of the stars. Its alignment with the solstice remains accurate to this day.

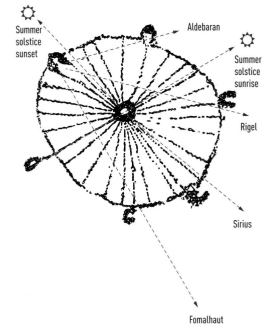

The Ensisheim Stone

Rocks from the sky

1492

◀ Fragment of the meteorite

In the ancient world, and even up to relatively recent times, no one imagined that rocks could rain down from the sky unless propelled by enemy artillery such as slings or catapults. Meteor showers would be observed, but rarely did one of these celestial visitors land near an observer of the event. Chinese records show, however, that rocks could not only fall from the sky in this way but could also be lethal and cause considerable damage. This realization entered the Western world's consciousness during a spectacular event a few minutes before noon on November 7, 1492, over the town of Ensisheim in France. A boy working in a nearby wheat field witnessed a 280-pound meteorite impacting the ground, digging itself a crater of more than three feet deep. In fact, the brilliant fireball and detonation were seen and heard more than one hundred miles from the town. It caused, to say the least, quite a sensation.

While its *original* mass was 280 pounds, the townspeople chopped off more than one hundred pounds to keep as souvenirs. Later, the "Thunderstone of Ensisheim" was brought into town and secured in the parish church with iron chains so that it would not wander about town at night.

The stone is the oldest meteorite whose appearance in the sky and impact on the ground can be identified with a precise date and time by a witness, and of which pieces are still preserved. It was also the first meteorite the news of whose fall was published in a number of broadsides and woodcuts after the invention of the printing press in the 1450s, resulting in a social media event of considerable impact covering three major towns near Ensisheim.

◀ 1492 leaflet bearing news of the meteorite

21

De Revolutionibus

Copernicus changes the center of the universe

1564

*D*e Revolutionibus Orbium Coelestium ("On the Revolutions of the Heavenly Spheres") is the fundamental work of the astronomer Nicolaus Copernicus, in which he describes his heliocentric theory. He began to write it around 1515 and finished it in 1531, although it was not published until the year of his death, in 1543. The Copernican model of the solar system was essentially the Ptolemaic geocentric model, but with the Sun stationary and Earth rotating upon its axis and revolving around the Sun. These dual motions create the perception from the surface of Earth that the Sun and planets orbit it. *De Rev*, as it's called in shorthand, provided an extensive mathematical discussion of how this "coordinate transformation" would work, but Copernicus's heliocentric model was never received as a work providing clarifying simplicity over the Ptolemaic model, as he had hoped it would.

The problem was that he, like others before him, had assumed the planets orbited upon exactly circular paths at a uniform speed. But in fact, as proved by Johannes Kepler about fifty years later, the orbits are ellipses, and planets do not move at constant speeds along their orbits. Copernicus's flawed understanding required him to use the established Ptolemaic idea that the planets were moving around a main circular orbit called the *deferent* while also traversing a smaller orbit called an *epicycle* to account for the variable speeds of the planets, which

affected his predictions. This erroneous heliocentric model produced the *Prutenic Tables*, a new ephemeris (table indicating an object's trajectory) of the predicted locations of the planets in the sky published in 1551 by Erasmus Reinhold. It in turn was eventually superseded by Kepler's more accurate *Rudolphine Tables* in 1627, based on elliptical planetary orbits without the epicycles.

It is hard to read modern astronomical textbooks without encountering some kind of debt owed to *De Rev*. While it didn't find immediate acceptance, due in large part to the Church's resistance, it did make its way to all the leading minds of the day. Harvard emeritus astronomer Owen Gingerich conducted a thirty-year worldwide campaign to inventory all of the currently surviving copies of *De Rev*. He ultimately found 276 copies of the first edition and 325 copies of the second. All of the major mathematicians and astronomers of the seventeenth century had their own copies. Moreover, many of them added marginal notes that helped Gingerich discern precisely which sections in *De Rev* were the most stimulating to its technical readers—apparently, those on planetary motion. By Gingerich's account, only the first edition of the Gutenberg Bible ca. 1454 has been researched and cataloged with the same level of detail.

net, in quo terram cum orbe lunari tanquam epicyclo contineri diximus. Quinto loco Venus nono mense reducitur. Sextum denique locum Mercurius tenet, octuaginta dierum spacio circum currens. In medio uero omnium residet Sol. Quis enim in hoc

pulcherrimo templo lampadem hanc in alio uel meliori loco po-
neret, quàm unde totum simul possit illuminare: Siquidem non
inepte quidam lucernam mundi, alij mentem, alij rectorem uo-
cant. Trimegistus uisibilem Deum, Sophoclis Electra intuentem
omnia. Ita profecto tanquam in solio regali Sol residens circum
agentem gubernat Astrorum familiam. Tellus quoque minime
fraudatur lunari ministerio, sed ut Aristoteles de animalibus
ait, maximã Luna cũ terra cognationem habet. Concipit interea à
Sole terra, & impregnatur annuo partu. Inuenimus igitur sub
hac

ordinatione admirandam mundi symmetriam, ac certũ har-
moniæ nexum motus & magnitudinis orbium: qualis alio mo-
do reperiri non potest. Hic enim licet animaduertere, non segni-
ter contemplanti, cur maior in Ioue progressus & regressus ap-
pareat, quàm in Saturno, & minor quàm in Marte: ac rursus maior in Venere quàm in Mercurio. Quodque frequentior appare-
at in Saturno talis reciprocatio, quàm in Ioue: rarior adhuc in
Marte, & in Venere, quàm in Mercurio. Præterea quòd Satur-
nus, Iupiter, & Mars acronycti propinquiores sint terræ, quàm
circa eorũ occultationem & apparitionem. Maxime uero Mars
pernox factus magnitudine Iouem æquare uidetur, colore dun-
taxat rutilo discretus: illic autem uix inter secundæ magnitudinis stellas inuenitur, sedula obseruatione sectantibus cognitus.
Quæ omnia ex eadem causa procedunt, quæ in telluris est mo-
tu. Quòd autem nihil eorum apparet in fixis, immensam illorũ
arguit celsitudinem, quæ faciat etiam annui motus orbem siue
eius imaginem ab oculis euanescere. Quoniã omne uisibile lon-
gitudinem distantiæ habet aliquam, ultra quam non amplius
spectatur, ut demonstratur in Opticis. Quòd enim à supremo
errantium Saturno ad fixarum sphæram adhuc plurimum in-
tersit, scintillantia illorum lumina demõstrant. Quo indicio maxime discernuntur à planetis, quodque inter mota & non mota,
maximam oportebat esse differentiam. Tanta nimirum est diui-
na hæc Opt. Max. fabrica.

De triplici motu telluris demonstratio.   Cap. XI.

Vm igitur mobilitati terrenæ tot tantaque errantium
syderum consentiant testimonia, iam ipsum motum
in summa exponemus, quatenus apparentia per ipsum tanquã hypotesim demonstrentur, quẽ triplicẽ
à Græcis uocari, dici noctisque circuitum proprium, circa axem
telluris, ab occasu in ortum uergentem, prout in diuersum mun-
dus ferri putatur, æquinoctialem circulum describendo, quem
nonnulli æquidialem dicunt, imitantes significationem Græco-

c ij   rum,

22

Tycho's Mural Quadrant

And his other tools of precision astronomy

1590

For centuries, since the time of Hipparchus and Ptolemy, accurate measurements of the locations of the stars in the sky were made infrequently, and few have survived to the present day, with the exception of Ptolemy's *Almagest* and the star catalogs of Han Dynasty astronomers. The accuracy of these star catalogs was only as good as the rather primitive diopters and theodolites available at the time. This level of precision was adequate for Ptolemy to create the first modern ephemeris of planetary positions, which endured until Copernicus authored his heliocentric model of the solar system in 1534. That all changed in the hands of Tycho Brahe, who in 1576 built extremely precise measurement instruments in his castle observatory, Uraniborg, on the Danish island of Hven.

Tycho recognized that the accuracy of the small portable instruments used by ancient astronomers could be vastly improved by simply making the instruments larger. Analysis of the thousands of star measurements among Tycho's logs suggests that he achieved an accuracy of about one-half arc minute for most of his instruments, providing a nearly tenfold improvement over the measurements reported in the ancient catalogs. One famous example was his mural (as in mounted on the wall) quadrant, a quarter-circle device he used to determine stars' altitudes.

The volume of high-quality data obtained was so enormous that Tycho hired the young Johannes Kepler in 1600 to organize everything and create what later became the *Rudolphine Tables*, published in 1627. But the subjects of Tycho's high-precision measurements included the planets as well. It had been Tycho's ambition to use them to prove his own model for the solar system, which was a hybrid between the Ptolemaic and Copernican models, but this was not to happen. A year after he hired Kepler, Tycho died, and it was left to Kepler to continue the planetary studies.

The high quality of the Tychonic planetary data enabled Kepler to uncover a variety of regularities in planetary motion now codified as Kepler's Three Laws. In addition, Tycho's data for his sightings of Mars led to the first dramatic change in our knowledge of planetary orbits. It was no longer possible to fit the data to a circular orbit. Instead the data revealed an ovoid or elliptical path around the Sun: Kepler's First Law. With this new fact, combined with his other two laws, Kepler fashioned an even more accurate ephemeris of planetary motions, the *Rudolphine Tables*. The tables were sufficiently accurate to predict a transit of Mercury observed by Pierre Gassendi in 1631 and a transit of Venus observed by Jeremiah Horrocks in 1639. Tycho's phenomenal measurement techniques formed the basis of astronomical predictions throughout the seventeenth century—but in 1690, the naked-eye 1,500-star catalog by Johannes Hevelius, and then John Flamsteed's telescopic catalog *Catalogus Britannicus* of 1725, superseded the Tychonic measurements.

23

Galileo's Telescope

The beginning of modern astronomy

1609

In *Starry Messenger*, Galileo shared detailed drawings of different phases of the Moon. His telescope's significant magnification provided evidence of the rocky nature of the Moon. ▶

For millions of years, the only human access to the details of the night sky came from the use of our eyes. These natural optical devices have a sensor (the retina) capable of discerning millions of colors and the arrival of individual photons of light. Our eyes' resolution is impressive, too; they can see at about the equivalent of a 576-megapixel camera.

But stargazing would be changed forever by the invention of the telescope, which essentially enhanced the basic faculties of the eye. Increasing the diameter of the eye's natural lens is biologically impossible, but in 1608, the Dutch eyeglass maker Hans Lippershey found a way to optically simulate the effect. He used a convex "objective" lens and a concave eye lens to create the first three-power optical instrument "for seeing things far away as if they were nearby," as he put it. Word of his Dutch perspective glass, as it was called, spread through Europe, inspiring Englishman Thomas Harriot to build a six-power telescope during the summer of 1609. This news then reached the ears of the Italian Galileo Galilei, who set about improving on the primitive device by grinding and polishing his own lenses until he reached a magnification power of about twenty-one times. Using his device, he was the first to actually see details in the night sky, beginning in 1609. His telescopes were unique in that they showed images right side up rather than upside down, which is common in simple optical instruments. With his superior instrument, he set up a side business of manufacturing and selling his "Galilean telescopes" to mariners.

Galileo would use his telescope to make seventy sketches of lunar details, phases of Venus, sunspots, star clusters, and the satellites of Jupiter that he published in his groundbreaking 1610 book *Sidereus Nuncius* ("Starry Messenger"). It was met with curiosity by some readers and derision by others, and Galileo's pronouncements about what he saw eventually led to his house arrest by the Vatican in 1633 on the grounds of heresy; the Roman Catholic Church was none too pleased by his claims that Jupiter's orbiting satellites proved the heliocentric model of Copernicus to be correct, not the Vatican-favored Earth-centered view. Not only was the unblemished disk of the Sun marked from time to time by moving spots, Galileo pronounced, but Jupiter was also an independent body in the solar system around which other objects orbited. This ran counter to the orthodox view that all things in creation revolved around Earth. Despite the Church's attempt to suppress his ideas, Galileo's view grew to be widely accepted. Thanks to his telescope, humanity's understanding of its place in the cosmos would never be the same.

Galileo's first refracting telescope ▶

TVBVM OPTICVM VIDES GALILAEI INVENTVM, ET OPVS, QVO SOLIS MACVLAS,
ET EXTIMOS LVNAE MONTES, ET IOVIS SATELLITES, ET NOVAM QVASI
RERVM VNIVERSITATE PRIMVS DISPEXIT. A. MDCIX.

24

The Slide Rule

The proto-calculator technology of the 1960s space program

1622

◀ A slide rule in action at the National Advisory Committee for Aeronautics Lewis Flight Propulsion Laboratory

John Napier—a Scottish landowner, mathematician, and astronomer—invented the logarithm, a mathematical function for performing calculations involving multiplication and division, and he presented it in considerable detail in his book *Mirifici Logarithmorum Canonis Descriptio*, published in 1614. Soon afterward, the English clergyman Edmund Gunter designed a ruler that could be used with the aid of two compasses to perform calculations in trigonometry using logarithms. The last step in the development of the slide rule was taken by the English clergyman and mathematician William Oughtred. In 1632 he devised an instrument consisting of two scales that slid past each other to perform multiplication and division.

The use of slide rules by engineers became commonplace in the 1800s, and the association of a slide rule with an engineer was as entwined in the public view as a stethoscope was for a physician. Although this particular technological marvel is known only to the truly gray-haired among us today, it enjoyed a bright, essential moment in the spotlight: The entire US space program through the last Moon landing depended on legions of engineers and scientists who turned to these manual calculators to solve engineering problems and literally take us to the Moon and back.

The slide rule came in many sizes, from six-inch miniatures to large circular devices made from a variety of woods and plastics. Based on the addition and subtraction of logarithms in place of multiplication and division of whole numbers, there were over a dozen different linear scales graduated in decimal values or in the trigonometric functions for rapidly performing advanced calculations with both large and small numbers. If you were a high school student in the 1950s and '60s, you carried your slide rule proudly into your physics and advanced math classes.

Pickett-brand slide rules were carried on Project Apollo space missions. The model N600-ES owned by Buzz Aldrin flew with him to the Moon on *Apollo 11* and was sold at auction in 2007 for $77,675. Once pocket-sized electronic computers became more readily available to engineers and scientists during the 1970s, slide-rule use began to diminish. The biggest changeover occurred once companies such as Texas Instruments and Hewlett-Packard began to market small electronic calculators that fit in a shirt pocket in the mid-1970s.

But today, some of us older scientists occasionally get nostalgic and dig through our attic boxes to locate our own pieces of history and remember when geeks ruled the world!

25

The Eyepiece Micrometer

The most precise astronomical measurement yet

1630

Double stars were known as far back as the time of the ancient Romans, when it is said that archers were selected based on whether they could discern the double stars Mizar and Alcor in the handle of the Big Dipper. But it wasn't until after the invention of the telescope that astronomers became more curious about double stars. Mizar's status as a double star was discovered around 1650, but it was many decades before astronomers paid serious attention to these systems. By 1718, only six double stars had been discovered, including the closest one, Alpha Centauri. The watershed event occurred when Reverend John Michell in 1767 announced that, following Newton's law of gravity, pairs of stars were actually in orbit around each other. From this, it would be possible to measure the mass ratios of the stars—a valuable and previously inaccessible measurement of objects beyond the solar system. Spurred on by this idea, a small catalog of known double stars was put together a decade later by Christian Mayer.

William Gascoigne was an English astronomer and instrument maker who was working on optical instruments in the late 1630s. By fortuitous accident a strand from a spider's web fell to exactly the right spot in the optical path of an instrument, and Gascoigne realized that he could use this together with a calibrated mechanical screw to make ultra-fine measurements. This became the forerunner of the engineer's micrometer. When he applied this same micrometer to the design of a telescope eyepiece, he was able to make precise measurements of the diameters of the Moon and planets.

Frederick William Herschel, using his large telescopes, began a study of double stars with a micrometer instrument of his own design. Within the eyepiece, he created a fiber that could be moved with a micrometer screw to measure the exact position of a star. Over the course of a year of study and measurement, instead of detecting the parallax shift due to Earth's motion around the Sun, he instead discovered that the stars themselves were moving in a curved path. Herschel's interpretation of this as the stars orbiting a common center of gravity was proved in 1828 by the French astronomer Félix Savary in his study of Xi Ursae Majoris. This discovery launched an intense phase of double-star cataloging and measurement that lasted until the advent of photography in the mid-1800s. Over one hundred thousand double stars are now known and have had their orbits measured; many of the early measurements involved the use of a micrometer.

◀ Gascoigne's micrometer as drawn by Robert Hooke

26

The Clock Drive

A new way of seeing with telescopes

1674

If you spend a little while gazing at the night sky, one of the things you'll notice is that the stars do not stay put. Over the course of hours, Earth rotates upon its axis, making a complete circuit every twenty-three hours, fifty-six minutes, and four seconds. That means that the view through a telescope will, maddeningly, show stars moving across the field of view. The idea of correcting for this apparent diurnal sky rotation was first applied to an armillary sphere in China by Su Song in 1094 using an ingenious water-clock mechanism. This technology was largely a novelty until the advent of large telescopes in the eighteenth century. English astronomer Robert Hooke wrote a paper about such a clockwork mechanism applied to telescopes in 1674. Soon after, in 1685, a clock-driven lens was designed by

Giovanni Cassini. The first telescope that actually used a clock-driven tracking mechanism was built in 1824 by Joseph von Fraunhofer, a master instrument maker. The roughly ten-inch "Great Dorpat" refractor at the Tartu Observatory in Estonia had an equatorial mounting and a clock drive that turned the polar axis to keep up with Earth's rotation.

Early clock drives were "powered" by falling weights, but even after the invention of the electric motor in 1834, motors powerful enough to turn the gear works of clock drives weren't available until the late 1800s. For most of the twentieth century, clock drives continued to be mechanical devices with numerous gears driven by motors. One of the major advances in clock-drive design occurred when computers became fast enough to keep up with the necessary calculations for tracking using a telescope mount that was non-equatorial "altitude-azimuth"

(Alt-Az; *altitude* is the distance of the object above the horizon; *azimuth* is the angular distance along that horizon). Virtually all modern telescopes larger than about nine feet now use Alt-Az mounts driven by computers and stepper motors, which continuously compute the correct azimuth and elevation for a target and update the telescope pointing several times a second or faster.

Thanks to these essential devices, spectroscopic and photographic studies were feasible for faint objects that required hours of exact guidance at the eyepiece to track. Without clock drives, either mechanical or electronic, most of the observations for which twentieth-century astronomy is famous, such as the discovery of the expanding universe and the photographing of surface details of distant planets, would have been impossible.

◀ This mechanical clock drive was used with the sixty-inch telescope at California's Mount Wilson Observatory until 1968, when it was replaced with stepping motors and electronic control systems.

27

The
Meridian Circle

An ingenious device to
help catalog the stars

ca. 1690

Long before GPS, navigation was performed by consulting a precision timepiece to determine longitude, and a sextant to determine latitude. For these tools to work, navigators had to consult something called an *ephemeris*: a collection of tables of the positions of stars and planets. For centuries, astronomers were obsessed with creating ever more elaborate catalogs of stars based on high-accuracy measurements of their positions in the sky.

The meridian circle, invented by Danish astronomer Ole Rømer around 1690, would help considerably in compiling these star catalogs. Affixed to a telescope, such as on the meridian telescope shown on the opposite page at the Kuffner Observatory in Vienna, the meridian circle was widely used in astronomical observatories and naval observatories during the 1800s to measure star positions and to observe star transits, allowing clocks to be set with extremely accurate times. These clocks were then used to set marine chronometers carried aboard ships to determine longitude, and as primary time standards before the invention of radio timekeeping via the short-wave WWV radio station and atomic clocks.

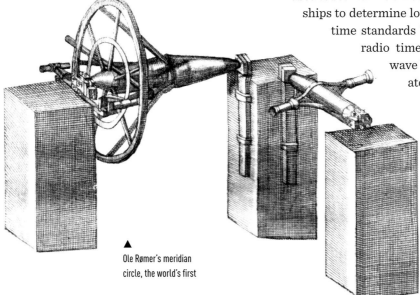

▲
Ole Rømer's meridian
circle, the world's first

The meridian (or transit) telescope was set up so that it could move only along the local north-south meridian line. The declination of a star (an astronomical coordinate similar to terrestrial latitude) was its location along this meridian line, which was measured to high accuracy using dials read with microscopes. Within the eyepiece, a precision reticle, made of sets of wires, gave an angular reference perpendicular to the meridian, which represented the star's right ascension coordinate (an astronomical coordinate similar to terrestrial longitude). If a star with a known right ascension passed exactly across the north-south line of the reticle, the precise local sidereal time could be calculated, and this could be used to correct the observatory's clock. Alternatively, if the star's right ascension wasn't known, you could use your local clock to record the exact time the star reached the meridian line in the eyepiece, in this way calculating the right ascension.

Using this process to create a precision star catalog was a monumentally tedious process requiring that you first identify the star of interest and then perform a transit observation to get its sky coordinates. By 1801, the best of these catalogs was published by the French astronomer Jérôme Lalande and contained over 47,000 stars brighter than magnitude 9.0. This was replaced by the Bonner Durchmusterung catalog of more than 320,000 stars, published from 1859 to 1862—the most comprehensive star catalog prior to the advent of photography. Thanks to transit observations, the positions of these stars could be known to fractions of an arc second.

A 19th-century meridian circle at the Kuffner observatory ▶

28

The Skidi Pawnee Star Chart

A relic of an American Indian tribe famed for celestial observation

1700

The Pawnee Nation of the Great Plains numbered more than sixty thousand people in the early 1700s and was one of the largest and most powerful tribes in the Plains region. The Skidi (or Wolf) Pawnee are a band of the Pawnee tribe who lived along the North Platte River in Nebraska. The Skidi band was well-versed in the study of the night sky. According to their belief system, the stars made them into families and villages and taught them how to live and perform their ceremonies. Their villages were even laid out according to a geometric plan that reflected the locations of certain important stars in the sky.

As with many other Native American tribes across North America, knowledge transfer historically relied on a variety of oral traditions, so written or artistic records that have survived the centuries are rare. Fortunately, some do remain, such as a soft leather buckskin measuring twenty-two inches by fifteen inches. It arrived at the Field Museum in Chicago in 1906 as part of a cache of objects acquired by anthropologists George Dorsey and James R. Murie. Murie was himself part Pawnee, and a detailed recorder of Pawnee customs. Now called the Skidi Pawnee Star Chart, it was included in Pawnee Sacred Bundle No. 71898, also called the Big Black Meteoric Star Bundle by the Pawnee. Although the precise age of the map is not known, it is believed to have been created sometime in the 1700s. The hide is covered with dots and crosses that mark the approximate locations of major stars and constellations, including a small group of six crosses representing the Pleiades, the circumpolar constellations Ursa Major and Ursa Minor, the North Star, called The Star That Does Not Walk Around by the Pawnee, and the Hyades star cluster in Taurus, shown just below the Pleiades as a V-shaped group of crosses. The circular shape of the Corona Borealis also appears to be present, along with a dappled collection of points running up the center of the map that likely represent the Milky Way. Altogether, it's a testament to the Skidi's exceptional powers of observation—and a beautiful work of art, too.

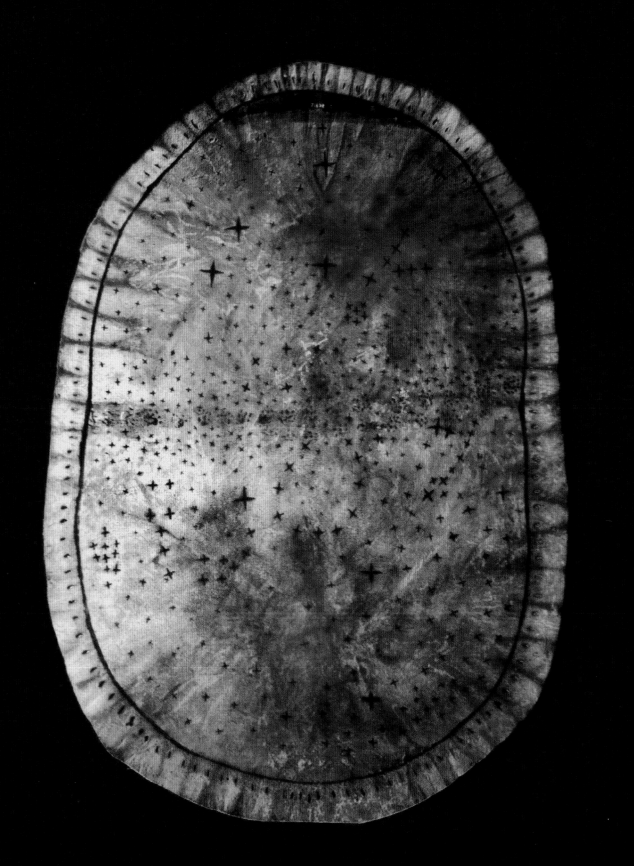

29

◀ Cover of *Harper's Weekly* depicting children viewing the 1882 transit of Venus through smoked glass

Modern eclipse-viewing glasses
▼

Smoked-Glass Sun Viewing

The original eclipse glasses, making celestial events accessible to the masses

1706

Although the history of viewing the Sun through smoked glass is mostly lost to time, there are nevertheless a few mentions of this inexpensive method in the historical literature. The earliest of these is probably in a short letter to the editor published in the prestigious *Philosophical Transactions of the Royal Society of London* in reference to the May 12, 1706, total solar eclipse.

The method is quite simple. When a piece of glass is tilted slightly above a candle flame, the soot from the combustion will eventually cover a large spot with enough density to greatly diminish the Sun's light.

Although professional Sun observers during the nineteenth century used specially made filters for safe observing, these resources were too costly for the average person, and the smoked-glass method was hugely popular during the total solar eclipses of the nineteenth and twentieth centuries. In some specific cases, such as the transits of Venus of 1874 and 1888, newspapers reported thousands of people using smoked glass to observe the tiny dot of Venus traversing the solar disk. As dangerous as this method was—while making viewing safer than it was with the naked eye, the glass didn't dim the Sun's luminosity enough that it couldn't damage the retina—it nevertheless made observation of solar eclipses massively accessible public events, stimulating public interest in astronomy.

Ophthalmologists began warning the public about the retinal damage that could be done by smoked-glass viewing in the late 1800s. During the eclipse of April 1912 over Europe, over 3,500 cases of retinal damage were reported in Germany. Such reports were also common after the November 12, 1947, eclipse, which passed over Los Angeles and resulted in dozens of children suffering from damaged and black-spotted vision.

Smoked glass was used through the 1940s, though the method began to lose favor when companies such as Harvey & Lewis Opticians in New England developed an Eclipse-o-scope for the August 31, 1932, total solar eclipse, which passed across Portland, Maine. The ten-cent cardboard viewer had two lenses made from dense film, and it was far more convenient and less messy than smoked glass.

Fast-forward to the transits of Venus in 2004 and 2012 and the total solar eclipse over North America on August 21, 2017, and you'll find solar viewing glasses not unlike the Eclipse-o-scope in fundamental design, but with filters to block harmful radiation. With the improved safety has come increased participation in such celestial viewing events. NASA and other institutions have distributed these modern glasses by the millions, and solar events since 2004 have been viewed by an estimated one billion people worldwide—now much more safely.

30

The Gyroscope

An ingenious device for keeping rockets flying straight and true

1743

Rockets suffer from a common problem. Even if you launch them straight upward, once in the air, they will reliably tilt over and crash due to sideways winds and forces. German scientists in 1934 found a solution to this problem in the gyroscope. A gyroscope is a mass that spins around an axis at high speed, producing a lot of angular momentum. When a force is applied to it to cause it to move, it resists that force and tries to maintain its rotation around the fixed axis. German scientists discovered that these whirling masses, like a strong invisible hand, could torque a rocket into a constant upright orientation as it traveled vertically. But the masses of larger rockets are too great for this brute-force approach to work. On March 28, 1935, American rocket scientist Robert Goddard

▲

Close-up of gyroscope
for Goddard's rocket

A V-2 rocket control
gyroscope ▶

successfully demonstrated a far better idea: Use three gyroscopes as the rocket's attitude (orientation) sensors.

No matter what the mass of the rocket, these gyro sensors could be tied to a system that controlled the jet vanes that directed the gases out of the rocket nozzle and kept the rocket in a perfectly upright attitude regardless of wind conditions. With the A-5 rocket launch, Goddard also showed that this system could be programmed to carefully tilt the rocket into horizontal flight, a maneuver needed to insert the rocket into an orbit. Because this approach was published in the open literature, Goddard's designs were eventually used by German engineer Wernher von Braun to develop the lethally successful V-2 rockets.

Just three weeks after Yuri Gagarin made his historic orbital manned flight into space atop a Redstone rocket, Alan Shepard became the first American astronaut to reach a suborbital altitude of 116.5 miles aboard a cramped Freedom 7 capsule on May 5, 1961. Neither of these successful but inherently dangerous flights would have been possible without an accurate inertial guidance system, all based on the simple concept of the gyroscope.

31

The
Electric Battery

Keeping spacecraft running

1748

Without electricity, it is safe to say none of the advancements in astronomy and space research would have been possible after the first few decades of the twentieth century. But establishing how "electricity" was discovered is a complex story that goes all the way back to ancient Greeks, including Thales, who noticed that rubbed amber attracted dust. However, it was not until 1745, with the advent of a prototypical capacitor, the Leyden jar (at that time, essentially a jar filled with water that had a wire dipping into it), that electric charges could be stored for more careful study.

A few years later, in 1748, Benjamin Franklin improved on the simple one-cell Leyden jar by combining several in a connected "battery" that yielded a single huge electrical discharge—large enough to be lethal. Various spinning mechanisms were designed to electrostatically charge these batteries, but there was no convenient way to create a sustained flow of charges until the revolutionary invention of the chemical battery by the Italian physicist Alessandro Volta in 1800.

Volta's battery was composed of alternating plates of zinc and copper separated by brine-soaked cloths. When a number of these cells were stacked up, a continuous electrical charge would flow through a wire connecting the bottom copper plate (the positive, or +, terminal, called the *cathode*) to the top zinc plate (the negative, or –, terminal, called the *anode*). With this combination, each cell generated 0.76 volts of electrical potential, so a two-thousand-cell pile such as the one constructed by Sir Humphry Davy in 1808 produced over 1,500 volts in the first demonstration of an arc lamp.

◀ Four Leyden jars joined into a simple battery

Batteries are essential for space exploration and astronomy. In space, electricity can be generated using solar panels or radioisotope (radioactive) thermoelectric generators (RTGs), but in many instances this electricity has to be stored for later use, especially when the Sun is eclipsed by a planetary shadow. During the early 1960s, spacecraft used nickel-cadmium (NiCad) batteries as part of their solar panel system. In the 1970s, more powerful lithium-based batteries were developed, and by the turn of the century, lithium-ion batteries began to dominate both in space and across the commercial market back on Earth, from portable power supplies to phones, laptops, and other devices. NASA originally used nickel-hydrogen batteries on the International Space Station, but these have been replaced by the higher-power lithium-ion batteries. For the Hubble Space Telescope, launched in 1990, which orbits Earth in approximately ninety-five minutes, about thirty-six minutes are in Earth's shadow. Prior to being replaced in 2009, its original six nickel-hydrogen batteries had been in use for eighteen years, delivering 450 amp-hours of electricity.

PREMIER VOYAGE AÉRIEN EXÉCUTÉ DANS UN AÉROSTAT À GAZ HYDROGÈNE
PAR **CHARLES** ET **ROBERT**, Le 1er Déc. 1783. DÉPART DES TUILERIES.

◀ An artist's postcard rendering of the world's first manned hydrogen balloon flight in 1783

Pilâtre de Rozier and d'Arlandes's Balloon

First flight

1783

Without a doubt, human entry into space during the last half of the twentieth century would not have been possible without the millennia-old interest in flight. When we think of early air travel, we might first think of planes and jets, starting with the Wright brothers' Kitty Hawk, North Carolina, first flight in 1903. But we actually broke the bonds of gravity and traveled through the air long before that. It all started with balloons.

The Montgolfier brothers, Joseph-Michel and Jacques-Étienne, following a series of experiments with bags lofted into the air with hot air from a fire, sent up into the skies on September 19, 1783, the very first hot-air balloon passengers ever: a sheep, a duck, and a rooster. The eight-minute flight reached an altitude of 1,500 feet and landed safely. Then, on November 21, 1783, the first manned flight on an untethered balloon took place, carrying Jean-François Pilâtre de Rozier and François Laurent le Vieux d'Arlandes to an altitude of 3,000 feet. Hot air was an extremely risky mode of transport—more than a few balloons caught fire during the early years of flight. In fact, the world's first aviation disaster involved a balloon: On May 10, 1785, the town of Tullamore, Ireland, was seriously damaged when a balloon crash started a fire that burned down about a hundred houses. However, Jacques-Alexandre-César Charles had another idea. Why not dispense with fire and use a gas such as hydrogen, which is lighter than air?

Brothers Anne-Jean and Nicolas-Louis Robert built the world's first hydrogen balloon for Charles, and it made its maiden flight on August 27, 1783. The first manned hydrogen balloon flight occurred soon after, on December 1, 1783. Wishing not to waste the scientific opportunity presented by this two-hour, 1,800-foot-altitude flight, they used a barometer and a thermometer to provide meteorological measurements of the atmosphere above Earth's surface. Charles himself later took a solo ride and reached an altitude of more than 6,000 feet.

Although balloons were used for studying the weather, the first scientific use of balloons to study the rest of the universe was by Austrian physicist Victor Hess. Between 1911 and 1913 he flew balloons to altitudes of several miles, carrying simple electroscopes to measure the electrical charging of the air at different altitudes. During a solar eclipse on April 17, 1912, he discovered that the air remained charged even without sunlight and concluded that the cause of the charging had to be something in space. Robert Millikan in 1928 found the source: cosmic rays. By the 1970s, balloons—now filled with helium—were routinely used for a variety of scientific observations at altitudes as high as one hundred thousand feet.

FIG. I.

33

◀ A drawing from *The Scientific Papers of Sir William Herschel*, published in London in 1912 by the Royal Society and the Royal Astronomical Society

William Herschel's Forty-Foot Telescope

The largest scientific tool of its day

1785

Before he began to entertain his passion for astronomy in the early 1770s, Frederick William Herschel was an accomplished musician, with twenty-four symphonies and many concertos and church organ compositions to his name. But when he set out to make his first telescope, a six-inch-diameter, seven-foot-long device, he was headed toward the findings that would leave the more lasting marks on history. He began an aggressive search and cataloging of double stars, which were fashionable objects of study during the post-Newtonian era. It was with this seven-foot telescope that, in March 1781, while searching for double stars, he noted a faint, moving object that would turn out to be the planet Uranus—the first planet discovered since antiquity.

Between 1782 and 1802, Herschel compiled a systematic catalog of 2,500 nonstellar objects using various telescopes and classified over 2,400 of them into specific morphological categories. Herschel's catalog, published in three parts in 1786, 1789, and 1802 as *The General Catalogue of Nebulae,* was later expanded by his sister, Caroline, and his son, John, to become the *New General Catalogue.* One of the most comprehensive records of deep-space objects available, an expanded and updated edition of it is still in use today. Virtually all of the bright nebulae and galaxies studied in the sky today have an NGC number in the catalog, such as the Orion Nebula, which is called NGC1973.

In 1785, Herschel built a forty-foot telescope under the patronage of King George III. It was the most massive scientific instrument of its day, with a speculum mirror (an alloy of copper and tin molded, ground, and polished into a reflective surface) roughly forty-eight inches in diameter, and a tube made of iron some forty feet in length. It was cumbersome to use and, according to Herschel, never gave as clear and focused an image as his smaller telescopes. Nevertheless, it was world-famous and a history-making object: At the time that Herschel constructed it, it was the largest scientific instrument ever made, and he used it to discover two additional moons of Saturn: Mimas and Enceladus. It represented the state-of-the-art in metal telescope mirror making, at least until William Parsons, the third Earl of Rosse, completed his seventy-two-inch-diameter "Leviathan of Parsonstown" in 1845.

34

The Spectroscope

Discovering what the stars are made of

1814

Sir Isaac Newton used light shining through a simple prism to experiment with the properties of sunlight, but it would take the technical genius of the German physicist Joseph von Fraunhofer to evolve this simple tool to its next stage in response to a very practical need. Fraunhofer was a brilliant optician and builder of exacting scientific instruments of his time. Working with brass and polished glass, he fabricated numerous scientific instruments for perfecting the precision manufacture of optical lenses for telescopes. But to carry his precision to the next decimal place, he needed a source of light at a single wavelength for ultra-precise lens making. This led him to develop prismatic and diffraction-based instruments to create pure light from sunlight. This need also inadvertently led him to discover one of the most useful features of sunlight and starlight: It carries information about the atoms that produce it.

Kirchhoff's spectroscope
▼

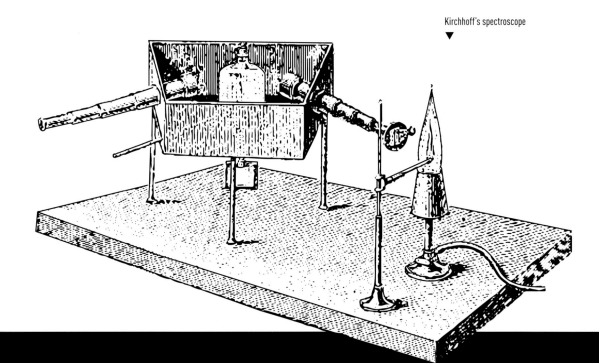

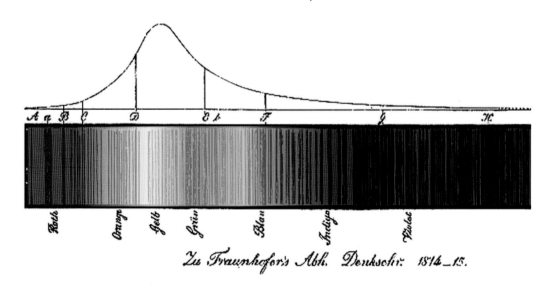

By 1814, it was known that refraction depended on the wavelength of light, but sunlight was a mixture of many different "colors." Fraunhofer experimented with prisms and telescopes to study dispersed sunlight at high magnification and invented a device that displayed these colors with great clarity: the spectroscope. His instrument was a theodolite telescope, like those used for surveying, but in Fraunhofer's tool the sunlight entered a prism first and then the telescope. Instead of using Newton's famous pinhole-in-a-window-shutter technique, he used a thin slit and discovered that the resulting spectrum had almost six hundred dark lines crossing it. The lines were in the same locations whether the sunlight came straight from the Sun or was reflected by the Moon first. Some of the lines even appeared in the same places as the lines in light from a flame from certain minerals being burned. The most prominent of these solar dark lines are now called *Fraunhofer lines*.

Gustav Kirchhoff and Robert Wilhelm Bunsen later discovered that what these lines indicated were the specific wavelengths of radiation from a light-emitting object that were being absorbed, revealing which elements—hydrogen, helium, and so on—were the sources of the light. This invention by Fraunhofer and his contemporaries revolutionized astronomy: Now you could determine the elemental composition of the Sun, stars, and any other astronomical matter that emitted light. Without this key technology, modern astronomy would still be stuck back in the eighteenth century, back when we had no idea what the universe is made of.

35

The Daguerreotype Camera

The advent of astronomical photography

1839

John Draper's first photo of the Moon ▶

Thousands of years ago, astronomers had to record what they saw in the sky through engravings and crude drawings, which often reflected the details of what the observer wanted to see rather that what was actually in plain view. Astronomical artistry reached a new level of skill and accuracy in the early nineteenth century but was rapidly overtaken by the arrival of photographic technology: the film camera.

The first of these cameras was invented by the Frenchman Louis-Jacques-Mandé Daguerre and introduced worldwide in 1839. Other photorecording techniques were being experimented with in the early 1800s, but the daguerreotype method soon distinguished itself. Daguerre did not actually patent and profit from his invention. Instead, it was arranged that the French government would acquire the rights in exchange for a lifetime pension. The government would then present the daguerreotype process free to the world as a gift, which it did on August 19, 1839. By 1853, an estimated three million daguerreotypes per year were being produced in the United States alone.

It didn't take long to turn the new invention toward the sky. In 1839, French physicist and mathematician François Arago delivered an address to the French Chamber of Deputies that set forth a long list of applications for photography; he included astronomy on the list. Daguerre himself had attempted the first known astronomical photograph earlier in 1839, but the resulting image was reportedly out of focus, and the photograph was later lost in a fire.

John William Draper, a professor of chemistry at New York University, managed to make the first successful, in-focus photograph of the Moon a year later, in March 1840, taking a twenty-minute-long daguerreotype image using a five-inch reflecting telescope. An 1845 daguerreotype by the French physicists Léon Foucault and Hippolyte Fizeau may be the first photograph of the Sun. The Italian physicist Gian Alessandro Majocchi made a failed first attempt to obtain a photograph of a total solar eclipse on July 8, 1842, in his home city of Milan. And during the transit of Venus on December 9, 1874, French scientist Pierre Jules César Janssen captured the motion of Venus crossing the Sun through a sequence of daguerreotype photographs.

Daguerreotypes continued to be used in astronomy until more advanced and less cumbersome film materials were developed in the 1870s. But it earned its place in space exploration long before it became obsolete, giving us our very first photographs of the heavens.

A daguerreotype camera from 1839 ▶

36

The Solar Panel

Fuel for spacecraft

1839

In 1839, the nineteen-year-old Frenchman Alexandre Becquerel was experimenting in his father's laboratory when he made a historic discovery: Silver chloride mixed into an acidic solution produced an electric current when exposed to sunlight. This became known as "the photovoltaic effect." In the 1870s, the photovoltaic effect was observed in selenium, a common and inexpensive byproduct of the processing of sulfur ores. Magic happens when you pour molten selenium onto a copper plate to make one electrode, then cover it with gold leaf to form a second electrode. When light shines on the semitransparent gold foil, it ejects electrons at the selenium-gold junction to create an electric current. More than just an exotic chemistry experiment, this discovery was the catalyst for what would become the green energy revolution a century and a half later.

Fritts's rooftop solar power system, from an old postcard ▶

◀ Solar panels on the International Space Station

Flash forward to 1883, when the thirty-four-year-old American inventor Charles Fritts developed the first solar cell. He rigged it up on his New York City rooftop, becoming the first person in history to try to produce large quantities of electricity from sunlight. Fritts's rooftop solar array converted only about 1 percent of the absorbed sunlight into electrical energy, but this watershed moment would pave the way for the steady improvement of solar-electric technology to come. Eventually, this technology would become instrumental to space travel. But first its efficiency had to be dramatically improved.

By 1941, Russell Ohl, an American engineer, had developed what's now considered the first modern solar cell, made with silicon. But it wasn't until 1954 that a solar cell capable of 6 percent efficiency was finally created. The advent of high-efficiency solar cells led to their use in consumer products, such as the first solar radio in 1957 (and later, solar calculators and watches in the 1970s). Now, solar power was ready to be used in space.

The technology didn't make it into the first US satellite, *Explorer I,* because that satellite was a hurried response to the Soviet Union's *Sputnik 1.* But under the lead of Dr. Hans Ziegler, a pioneer in the use of solar power in space, solar panels were incorporated into the design of *Vanguard 1.* When it was launched in 1958, its solar panels became the first ever to be used on a spacecraft.

Solar power remains the workhorse generator of electricity for satellite systems today. Solar panels are used in space for two reasons: to operate the gadgets and gizmos on board, including heating and cooling, and for propulsion. They're so critical to spacecraft operation that they're generally built to be able to pivot, so they can always stay in the most direct path of light.

The largest collection of solar panels in space today is, unsurprisingly, that of the International Space Station, which uses 262,400 solar cells, together more than half the size of a football field, to generate up to 120 kilowatts—more than enough power to run its systems.

37

The Leviathan of Parsonstown

The last telescope of its kind

1845

William Parsons's ca. 1845 drawing of Messier 51, the Whirlpool Nebula
▼

William Parsons, the third Earl of Rosse, was a wealthy astronomer who had inherited his father's estate in County Offaly, Ireland. A graduate of Oxford University with a degree in mathematics, his interests in astronomy led him to fabricate a number of telescopes having mirrors of speculum, which is an alloy of copper and tin. He eventually formulated a research plan: Prove (or disprove!) Immanuel Kant's 1755 hypothesis that planetary systems formed from gravitationally collapsing, swirling disks of gas by finding examples of this among the many nebulae cataloged by William Herschel. To do so, he would have to build a telescope large enough to see faint details clearly within the otherwise indistinct shapes of the nebulae.

Parsons's challenge was that no one had ever constructed a Newtonian reflecting telescope with a six-foot-diameter speculum mirror. No instruction manual existed, and no one was interested in sharing the tricks of the trade for grinding and figuring (polishing the mirror to a perfect optical shape) such a mammoth three-ton mirror. His efforts began in 1842, and after considerable effort on the part of a construction crew, the so-called Leviathan of Parsonstown was completed in 1845. But then the Irish Potato Famine struck, pulling Parsons away from astronomy to offer funds and assistance to those in need. At the tail end of the famine, observations commenced in 1848 and resulted in Parsons's observations and drawings of Messier 51, the Whirlpool Galaxy, and Messier 1, the Crab Nebula—his first studies. The optics were sufficient to see the details in the spiral arms of Messier 51, which has become Parsons's trademark discovery. The Leviathan was used for research until 1890, but its aperture was not surpassed until the construction of the one-hundred-inch Hooker telescope at the Mount Wilson Observatory in Southern California in 1917.

The Leviathan played a historic role in the engineering and optics of large telescopes. It had the last large speculum mirror, having proved that at a certain size, using metal for the mirror caused too many problems, including that speculum is hard to shape, and it tarnishes easily. This led to the search for other materials. By 1856, Carl von Steinheil and Léon Foucault had devised a process that deposited a thin film of silver on a block of glass. In 1879, Andrew Common fabricated the first silvered-glass telescope mirror of three-foot diameter, launching a steady stream of progressively larger single-mirror telescopes, such as the Hooker one-hundred-inch at the Mount Wilson Observatory in 1917 and the Hale two-hundred-inch at the Palomar Observatory in 1948.

Speculum mirrors in reflector telescopes remained the only practical telescope components for over two hundred years, from the time of Newton in 1668 to the Leviathan of Parsonstown fell into disuse in 1890. As other technology supplanted it, the Leviathan has come to represent the end of an era.

Crookes tube with concave cathode, designed by William Crookes himself

A modern Crookes tube, glowing green with cathode rays

38

Crookes Tube

Detecting and measuring nuclear particles

1869

A stronomers rely on the mass spectrometer to help them identify the mass of any given particle, whether it's a cosmic ray particle or a particle trapped in the radiation belts of one of the planets. It makes a big difference to our theoretical understanding of the universe if a particle is an ordinary and common hydrogen atom, say, or an exotic iron or uranium atom. Many of the scientific advances in our understanding of the universe would never have been possible if astronomers had been unable to make these distinctions.

British physicist William Crookes experimented with a variety of discharge tubes between 1869 and 1880. These partial-vacuum tubes had a metal plate at one end (the cathode), and a second plate at the other end (the anode). A battery was placed between these

electrodes, producing a fluorescent green glow in the gas filling the tube. Then the anode, which was shaped like a Maltese cross, would project a shadow on the glass wall of the tube behind it.

A great many experiments were conducted during the late 1800s to discover what the green glow—the cathode rays—were. In one version of the tube, the anode was replaced by a disk with a hole in its center so that a beam of these cathode particles could be created and seen within the interior of the tube. In 1897, Crookes then placed a magnet across the beam and discovered that the beam's path would be deflected upward or downward depending on the orientation of the magnet's north and south poles. This simple experiment, along with others, eventually revealed that cathode ray particles were simply electrons.

While investigating ionized neon atoms using a similar device, British physicist J. J. Thomson and his assistant, Francis Aston, discovered that the resulting deflections of the atoms differed and produced two spots. They had confirmed that neon comes in two different forms: one with a mass of twenty-two and one with an atomic mass of twenty. The heavier neon, called *meta-neon* at the time, was an isotope of neon, which we now know has two additional neutrons. Aston continued these isotope investigations using this technique and built a new device, which he called a *mass spectrograph*. He quickly tested a variety of other elements and found that they also had several isotopic forms. Virtually all of the investigations into isotopes were conducted by Aston, who discovered more than two hundred naturally occurring isotopes. He went on to receive the Nobel Prize in Chemistry in 1922 for his discovery of isotopes using mass spectrometry. But it all began with the Crookes tube, invented a half-century before.

Aston's instrument is a workhorse in space research. Mass spectrometers are carried on virtually all spacecraft to quantify the particles in the solar wind, the particles in the radiation belts of Earth and the planets, and the composition of planetary atmospheres; they are also used to identify the nature of cosmic rays.

▲
The De Forest Audion tube from 1908

39

The Triode Vacuum Tube

The birth of electronics

1906

Although Guglielmo Marconi invented the first Hertzian wave wireless telegraphy system in 1894, it would take another twelve years before these weak signals could be amplified by using American engineer Lee De Forest's invention of the Audion vacuum tube in 1906. Before the advent of the vacuum tube, specifically the triode, the changing electrical currents from Marconi radio signals were fed directly into earphones to electromagnetically move a sound-producing diaphragm. Technological improvements were focused on making progressively more sensitive headsets for listening. The invention of the triode vacuum tube dramatically changed the development of radio-wave receivers by improving the actual strength of the currents feeding the headphones.

De Forest's triode tube (Patent No. 879,532) was only slightly more complicated than an Edison electric light bulb: It had a filament and a plate. The filament was heated by a battery current, and as the electrons were ejected from the filament, they traveled through a vacuum to the plate so that a current existed within the tube. De Forest placed a grid of wires between the filament and the plate (hence a triode) and discovered that by changing the voltage on the grid wires, he could control the filament-plate current. In fact, the current applied to the grid circuit was far weaker than the filament-plate current, so this arrangement actually amplified the weaker signal in the plate circuit. This amplification was later applied by Edwin Armstrong to the design of the first regenerative radio receiver in 1912, allowing for practical communication using the radio, which propelled the technology into popular use.

That's where space exploration comes in. It's impossible to pick a single foundational object that unlocked the potential for the space age, because all technology is derived from developments that came before. But the triode tube may come nearer than most: It's often considered the starting point of the existence of electronics. And the sending of electrical signals is fundamental to exploring space: A spacecraft that cannot communicate over long distances simply won't work, and it was the triode and the subsequent transistor technologies it spawned that made such communication possible. This process of electronic amplification of weak signals is also instrumental in today's detection of radio signals from the distant cosmos, helping us map the skies and search for extraterrestrial life.

40

The Ion Rocket Engine

Game-changing propulsion

1906

Even the name sounds exotic: ion rocket. Yet the basic technology behind them is simple and elegant, and in fact for decades you had one of these in your living room and didn't even know it—the television. It had been known since the late 1800s that a beam of electrons ejected from a filament to strike a phosphor screen would cause the screen to glow. The basic operation of an old-style cathode-ray television picture tube controlled this beam to paint a picture. The electrons slamming into the screen at nineteen thousand miles per hour transported momentum to the glass, but this was dissipated due to the much larger inertia of the picture tube—so none of the components of the television physically moved. However, if you took the same cathode ray and placed it in a vacuum chamber with the filament free to move, it would be pushed in the direction opposite to the screen, just like a rocket.

This idea—that cathode rays could replace chemical propellant—had been known as early as 1906, when Robert Goddard wrote about this possibility in his laboratory notebook. In fact, Goddard spent almost as much time developing the principles for "ion rockets" as he had for his liquid-fueled chemical rockets. In 1920 he even took out a patent on a "Method of and Means for Producing Electrified Jets of Gas."

The entire concept of ion rocket motors was greatly expanded theoretically and turned into practical rocket designs by the German physicist Ernst Stuhlinger, who later worked on V-2 rockets with Wernher von Braun. Once World War II came to an end and virtually all of von Braun's rocket team emigrated to the United States, their work on ion engines continued, but it wasn't until 1961 that, through NASA, the first test of this technology was attempted.

NASA began to investigate cesium- and mercury-fueled engines using Stuhlinger's designs, with the first engine operating at two thousand watts tested on September 27, 1961. This led to the launch of the first ion engines on a satellite, *SERT-1*, in 1964. While one engine didn't work, the other operated under constant thrust for half an hour, not only returning valuable data but also proving that ion thrust would work in the vacuum of space. During the heyday of commercial satellite launches starting in the 1980s, ion thrusters were installed on many satellites to provide the small, gentle "station-keeping" shoves needed to keep a geosynchronous (having an orbit fixed to a particular location on Earth) satellite within its assigned orbital slot. But engineers continued

to push the technology by using higher power and more complex engine designs to overcome a variety of technical problems.

In 1998, NASA's *Deep Space 1* (DS1) became the first spacecraft to use ion propulsion as its primary propulsion source. Its engine ejected a stream of xenon ions and provided a constant thrust of 0.09 newtons over the course of 16,000 hours. The thrust needed to alter the spacecraft's solar orbit to reach the asteroid 9969 Braille and the comet 19P/Borrelly used only about 330 pounds of xenon. Other spacecraft soon followed DS1's technological lead, such as *Hayabusa* (Japan, 2003), *SMART-1* (European Union, 2003), *Dawn* (United States, 2007), and *BepiColombo* (European Union and

Japan, 2018). Meanwhile, engineers continue working to increase the thrust of ion engines, with the current record holder, the X3 "Mars Engine," having been tested in 2017 at a thrust of 5.4 newtons.

A test of the X3 ion engine
▼

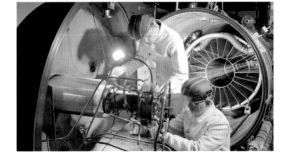

41

◀ The Hooker one-hundred-inch reflecting telescope at the Mount Wilson Observatory

The Hooker Telescope

The most famous telescope

1917

With thousands of powerful telescopes operational worldwide, and many more having been decommissioned and dismantled, it's a fool's errand to try to pinpoint just one that would be considered the most important to space history. Instead, we'll go with the one that's arguably the most famous: the Hooker telescope on Mount Wilson in California. Its massive size, including a one-hundred-inch mirror, unprecedented for its era, required all kinds of monumental support.

For a sum of $45,000 (more than $1 million in today's dollars) the American ironmaster and amateur astronomer John D. Hooker supported the grinding and polishing of the one-hundred-inch mirror. Andrew Carnegie, meanwhile, provided the balance of the funding to build the telescope and dome. But getting the funds to build it, which was orchestrated by George Hale, who already had the sixty-inch Hale telescope to his credit, was actually the least of the trouble.

The nine-mile road leading to the summit of Mount Wilson had to be widened, and the huge mirror blank had to be made. It was ordered from a glass factory in Saint-Gobain, France, in 1906, and was delivered in 1908; it then took five years of grinding and polishing to transform the four-ton glass disk into a mirror. The most daunting challenge was the design of the clock drive, which kept the telescope pointed at a star as Earth rotated. The massive two-ton clock gears were driven by a two-ton falling weight, and yet, when finally put in use, the drive had to work as well as a fine Swiss watch.

From its completion in 1917 until 1949, the Hooker remained the largest telescope in the world, and during its tenure it managed to attract some of the most exciting research of that era. The telescope was outfitted with a stellar interferometer in 1919 and succeeded in making the first measurements of the diameter of a star (Betelgeuse). In 1923, Edwin Hubble used the Hooker to detect variable stars in the Andromeda galaxy, for the first time proving that these nebulae were outside the Milky Way. Then in the late 1920s, Hubble and his colleague Milton Humason measured the speeds of dozens of galaxies, demonstrating Hubble's law and thus confirming that the universe is expanding.

42

◀ Goddard posed next to the rocket as it stood in the frame before being launched.

Robert Goddard's Rocket

The first use of liquid rocket fuel

March 16, 1926

Although there is some dispute over who came up with the idea of using liquid rather than solid fuel for rockets, the first demonstration of its use in an actual rocket occurred on March 16, 1926, in Auburn, Massachusetts, when Robert Goddard launched a rocket using liquid oxygen and gasoline as propellants. He would later write in his journal: "It looked almost magical as it rose, without any appreciably greater noise or flame, as if it said, 'I've been here long enough; I think I'll be going somewhere else, if you don't mind.'" Goddard had unleashed a technology that would be responsible for the colossal rockets to come.

Liquid fuel packs more of a punch than solid fuel, and it's easier to precisely start and stop its burning on command, opening the door to bigger rockets and more advanced maneuvers. German rocket engineers would take the first steps toward that bigger, more powerful future as they perfected their liquid-fueled V-2 rockets in the 1930s, which in turn would pave the way for the famous liquid-fueled launch of *Sputnik 1* in 1957 (see page 122). And because of the greater control with liquid fuel compared to solid, it's safer, and therefore it became the preferred fuel for missions with humans on board, including NASA's Mercury, Gemini, and Apollo missions, the latter of which required the increased thrust of liquid fuel to lift its tons of payload into the sky.

Liquid-fueled rockets remain the workhorses of twenty-first-century space exploration, from the launches of scientific payloads to Mars to the new rocket engines of the private company SpaceX, with its *Merlin 1D* (205,500 pounds of thrust) and *Blue Origin BE-4* (550,000 pounds of thrust). The latest NASA-designed engine, the J-2X, to be used on the *Orion* launch vehicle (replacing the now-retired shuttle), uses a liquid-hydrogen and liquid-oxygen mixture to provide 294,000 pounds of thrust.

Goddard's rocket may not have traveled very far or generated any earthshaking noise or enormous flames, but it left an enormous mark on space exploration, forever changing the way we send things up beyond the skies.

◀ The J-2X engine being prepared for testing in Mississippi. Each engine weighs more than two and a half tons and is expected to produce about 25 percent more thrust than the original J-2 engine used on the *Saturn V* Moon rockets.

43

The Van de Graaff Generator

The dawn of atom-smashing technology

1929

Cutaway drawing of the Westinghouse Atom Smasher, housing a huge Van de Graaff generator (note the two vertical fabric belts) ▶

Astronomy relies on an accurate understanding of the nature of matter and how it interacts over time and space. Since the early years of the twentieth century, this knowledge has steadily improved thanks to the design of powerful laboratories called *particle accelerators*, more popularly known as atom smashers, where subatomic particles are accelerated to very high speeds and then slammed into atoms, as well as other particles, to see what gets dislodged. You might think that only protons and neutrons would be ejected, but in fact, in keeping with Albert Einstein's formula $E=mc^2$, the energy produced in a collision not only shakes loose subatomic particles but also creates them. Furthermore, the higher the energy of the particles, the smaller their wavelength, thanks to the basic rules of quantum mechanics. This means you can use the colliding particles to see finer and finer details, much like the light in a microscope.

But these high-speed collisions would never be possible without the work of American physicist Robert Van de Graaff. In 1929, while working at Princeton University, he invented an ingenious device for accelerating particles to high energy. Basic electrical principles say that if you increase the voltage difference between two points in a conducting wire, the current in the wire will flow faster. This principle also accounts for lightning bolts, which occur when the voltage difference between a cloud and the ground is increased through friction. A Van de Graaff generator works by using friction to create charges—static electricity—in a circulating band of cloth. The charges are collected on a sphere isolated from the ground by insulators. As more charge accumulates, the voltage difference between the sphere and the ground steadily increases. This voltage difference can be used to accelerate other charged particles and focus them on a target to make a collider.

The first machine Van de Graaff designed to test this principle was made of an ordinary tin can, a small motor, and a silk ribbon. After receiving more funding for the project, he was able to make an improved version, and by 1931, he could report achieving 1.5 million volts, and he remarked that "the machine is simple, inexpensive, and portable. An ordinary lamp socket provides the only power needed."

Sparks thrown from one of Van de Graaff's own generators at MIT in 1933 ▶

In 1937, the Westinghouse Electric Corporation built a particle accelerator using an enormous Van de Graaff generator to explore the practical applications of nuclear science for industrial purposes. It was located in Forest Hills, Pennsylvania, and stood sixty-five feet tall. Two fabric belts traveled up the forty-seven-foot shaft to the collecting sphere, and the entire system was contained inside a pear-shaped enclosure filled with air pressurized at 120 psi to keep charges from leaking off the surface of the sphere into the atmosphere. Inside the shaft, between the belts, was a long vacuum tube through which the charged particles flowed to the collision target at the base of the pipe. The energy that the particles reached was simply the voltage difference achieved by the collecting sphere. The longer you let the belts run, the more charge was collected and the higher the voltage reached, allowing the accelerator to achieve high enough energies that it revolutionized the study of nuclear energy—a crucial, reliable power source for today's spacecraft and a window into the nature of cosmic matter that makes up the stars and galaxies in our universe.

44

The Coronagraph

Eclipses on demand

1931

For centuries, astronomers have been given tantalizing evidence that the Sun is surrounded by its own atmosphere, the corona, and that there are a variety of small features around the limb of the Sun that come and go in time. But for the same reasons that you turn off interior lights while you drive so as not to obscure the road ahead, it's only through the careful study of total solar eclipses that these features have been revealed: They are a million times fainter than the disk of the Sun, so normally they're overpowered by the Sun's light and can't be seen. During an eclipse, the disk of the Moon blocks the intense sunlight so that these faint features can be easily sketched or photographed.

In 1931, the French astronomer Bernard Lyot, of the Meudon Observatory, came up with a revolutionary new way to use this principle to develop a telescope's power of vision when aimed near the Sun, essentially by making artificial eclipses. The basic idea is rather simple: Within the telescope, place a black disk the same size as the Sun over the image of the Sun's face to reduce the light. In practice, however, the method was challenging. It wasn't easy to determine where along the optical path from the mirror to the eyepiece one should place this occulting mask. Lyot worked on this problem, eventually not only determining the correct placement of the occulting mask but also adding an invention of his own: light blocks, called *Lyot stops*, which remove scattered light from the Sun.

Lyot's ingenious system works like this: Light enters the telescope and is brought into focus by a lens. Instead of placing the camera at this spot, an occulting mask is used that exactly matches the diameter of the Sun's image at the point of focus. The surrounding light from the corona can then pass over the mask. But there is a problem: The mask produces a diffraction ring around the Sun that lets stray light from the Sun's disk mix with the faint coronal light. This is where the optics get complicated. The masked image then has to be reimaged by a second lens, and a second mask has to be inserted to block the stray light from the first image. Then a third lens must be added to bring the new image into focus. Now all you have is a black disk perfectly covering the Sun and no diffracted light, leaving behind only the faint light from the corona.

While the technology has evolved, the original concepts behind the coronagraph have provided a wealth of incredible insight. In the late 1990s, the NASA–European Space Agency joint-venture coronagraph on the Solar and Heliospheric Observatory spacecraft provided scientists and news networks with dramatic images of plasma ejecting from the Sun during major solar-storm events. For the first time, solar storms and other related events, collectively called *space*

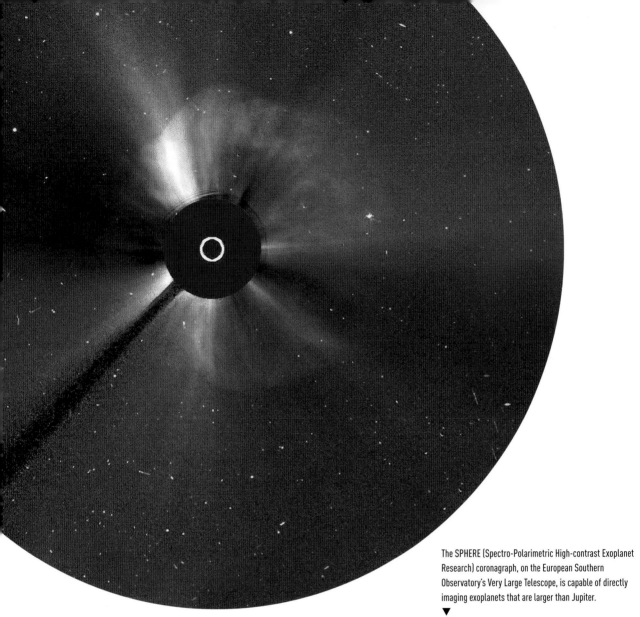

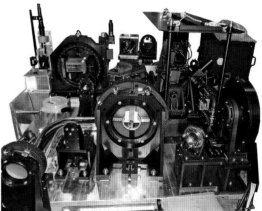

The SPHERE (Spectro-Polarimetric High-contrast Exoplanet Research) coronagraph, on the European Southern Observatory's Very Large Telescope, is capable of directly imaging exoplanets that are larger than Jupiter.

▼

weather, became a talking point during the nightly news—especially when it caused satellites to go out of commission.

Coronagraphs have become very popular on ground-based solar telescopes, in spacecraft observatories, and even in the search for extrasolar planets, since bright starlight can be eliminated and the faint light from closer planets can now be discerned.

▲

Reconstruction of Karl Jansky's first radio telescope from 1932 located at the National Radio Astronomy Observatory in Green Bank, West Virginia

45

Jansky's Merry-Go-Round Radio Telescope

The birth of radio astronomy

1932

By the early 1930s, radio technology had taken the commercial airwaves by storm, with radio receivers in nearly every home and elaborate radio station programming that fostered family listening hours. A few decades before that, as early as 1896, the German astronomers Johannes Wilsing and Julius Scheiner had theorized that there might also be natural radio waves from cosmic sources that could make their way to us. But the conclusion was that the ionosphere, a layer of Earth's upper atmosphere, would deflect these waves back into space before they reached Earth.

Flash forward to 1931. Karl Jansky, a radio engineer with Bell Telephone Laboratories, was trying to track down the source of noise in transatlantic radio transmissions. He built a large radio antenna that was dubbed Jansky's merry-go-round because its

orientation could be changed on a revolving platform. Now considered the first radio telescope, the device detected a type of radiation called *radio waves*, and over the course of a year Jansky collected volumes of data on an analog pen-and-paper recorder. The amplified signal from the antenna moved the pen to record the ups and downs of the signal's intensity.

One of the first things Jansky noticed was a strong signal that came and went every twenty-four hours, which led him to believe he was detecting radiation from the Sun. But there are two ways of measuring the day: The solar day, measured by Earth's rotation relative to the Sun, which takes twenty-four hours, and the sidereal day, determined by its rotation relative to the stars, which, because Earth constantly revolves around the Sun, takes a little longer—about four minutes longer. This was how Jansky eliminated the Sun as the culprit: The timing of the signal matched not the movement of the Sun across the sky and through the beam of his antenna but instead, the twenty-three hours and fifty-six minutes of a sidereal day. This led the engineer to the idea that he was detecting a radio source from outside the solar system. He eventually deduced that the signal corresponded with the center of the Milky Way, near the constellation Sagittarius. Until then, no one had ever proved that the cosmos emitted radio waves. Designating the source Sagittarius-A, the discovery became an instant front-page sensation, appearing in the May 5, 1933, edition of *The New York Times* under the headline: New Radio Waves Traced to Centre of the Milky Way.

This marked the beginning of radio astronomy, a new way of exploring the universe based on the detection of cosmic radiation. At first, though, few other astronomers knew what to make of this strange new frontier of research. But Jansky's discovery did catch the attention of radio amateur Grote Reber, who single-handedly built his own thirty-two-foot parabolic antenna in 1937 and compiled the first sky map in the radio spectrum. Jansky and Reber's instruments were handmade and rudimentary, but the basic principles of their inventions went on to enable dramatic discoveries as technology improved, including, decades later, the momentous detection of the fireball of radiation from the Big Bang.

46

The V-2 Rocket

The first artificial object in space

1942

So many things have to go exactly right for a successful rocket launch that many scientists say it still amazes them to see a rocket leave the ground without malfunctioning. Every single liftoff that doesn't nosedive or explode or fail to launch owes its smooth trajectory to the efforts of German rocket scientists led by Wernher von Braun in the 1940s. Their work on the V-2 rocket program, in which they overcame a string of failures and explosions on the launchpad, sorted out the engineering details of liquid-fueled rocket engines. Prior to the invention of liquid-fueled propulsion by American physicist Robert Goddard in the 1930s, rockets were still basically solid-fueled systems no more advanced than the gunpowder firecrackers developed thousands of years before by Chinese inventors. The efforts of the German scientists, engineers, and thousands of concentration camp slave laborers changed everything, leading to one of the deadliest battlefield weapons ever invented. It also happened to usher in the Space Age.

It all began on October 3, 1942, when von Braun and his team, at a classified launch site at the northeast tip of Germany, sent a V-2 missile about fifty-six miles (ninety kilometers) into space. This rocket is widely considered the first human-made object to reach space. That day, von Braun's boss declared: "This afternoon the spaceship was born."

Later, scientists more clearly defined the altitude where space begins as the Kármán line, one hundred kilometers (sixty-two miles) above Earth's surface. That threshold was broken by a V-2 rocket, too, on June 20, 1944.

The V-2 could be considered the creator of not only the Space Age but also the Space Race, as the Allied forces scrambled to appropriate the scientists and technology behind the program after Germany's defeat in World War II. While the Soviets developed their own missile program, the Americans embraced the immigrant von Braun, who continued his work on intercontinental ballistic missiles as part of the Redstone Program in Huntsville, Alabama. One of the first successful launches by American engineers of a V-2 into space came July 24, 1950, with the launch from Cape Canaveral of the *Bumper 8* two-stage rocket. It reached an altitude of 10 miles and traveled 160 miles (257 kilometers) from its launch, and carried simple instruments for measuring atmospheric temperature, pressure, and the newly discovered "cosmic rays" ground-based physicists had detected. But it was the successful launch of *Sputnik 1* in 1957 by the Soviets that would change rocket research from a military and scientific endeavor to a program focused on regaining geopolitical advantage through dominance of the final frontier.

◀ V-2 rocket engine

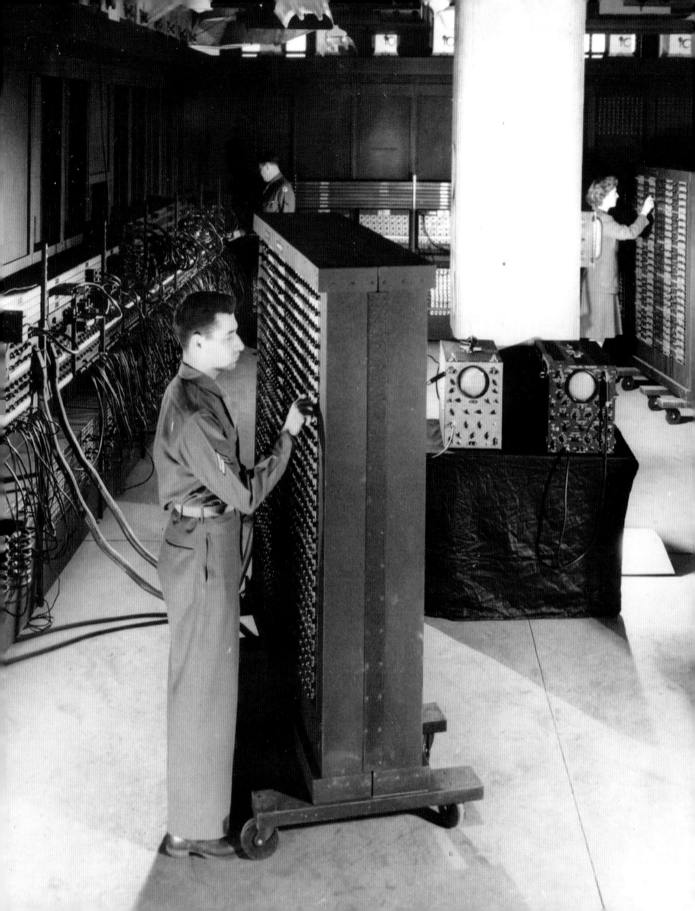

47

ENIAC

The first modern computer

1943

The modern idea of an electronic computer replacing mechanical calculators became possible once the Fleming valve, a vacuum tube, was invented in 1904 by Englishman John Fleming. Although used in early radio receivers, these vacuum tubes could actually be used as fast switches to turn current flows on and off within milliseconds. This switching ability is at the heart of all computers that use binary coding to store and process information. The first all-electronic computer to use vacuum tubes was built between 1937 and 1942 at Iowa State University by John Atanasoff and Clifford Berry.

The first modern computer designed with a memory, program storage, and execution modules began to be built in 1943 at the University of Pennsylvania and was called the Electronic Numerical Integrator and Calculator (ENIAC). In 1946, after World War II, it earned the nickname of the Giant Brain. In thirty seconds, it could calculate a complex missile trajectory that would take a human over twenty hours. By 1956, at the end of its life, it had grown to over twenty thousand vacuum tubes and five million soldered electrical connections, and

it weighed thirty tons. To store a ten-digit number required a bank of 360 vacuum tubes—likely the inspiration for the term *memory bank*, which came into popular usage by the 1960s. It's been said that the vacuum tubes of these early computers were so warm that insects were sometimes attracted to them (which might actually be urban legend, although at least one moth was confirmed within the Mark 2 computer), leading to the phrase "computer bug," or "bug in the program." ENIAC used so much electrical power that some rumors claimed lights actually dimmed in Philadelphia each time the computer was operated.

The image below shows an eight-tube logic module—a representative component of the early vacuum tube computers. Each tube had a cathode, which produced electrons, and an anode, which collected the electrons to create a current. In between was a grid of wires whose activation could either permit (represented as 1) or block (represented as 0) the flow of the current, representing the binary math used in a computer.

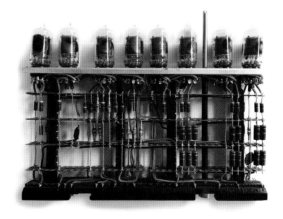

48

Colossus Mark 2

First programmable computer

1944

During World War II, British code breakers grappled with the Lorenz cipher, ultimately finding a way to decipher teletype messages encrypted in this complex German code. Tommy Flowers, a senior electrical engineer at the Post Office Research Station, was brought in to devise a better approach to one of the stages of the decryption process that, at that time, involved a purely electromechanical device that frequently failed to work at the required speeds. Flowers decided to use vacuum tubes rather than gears to replace the parts in the electromechanical machine. His first functioning design was the prototype Mark 1, nicknamed "Colossus" by the cryptanalysts working on the Lorenz cipher at Bletchley Park in 1944. The Mark 2 was operational on June 1, 1944, and consisted of 2,400 vacuum tubes, a tape transport using photocells to read the encrypted data, and switches to program which decryption function to apply. This feature made it the first computer of its kind to be programmable rather than built to perform only one kind of function.

The Colossus project was so top secret, its existence was not revealed until the 1970s. Flowers had been ordered to burn all records and notes of the computer's design and existence, and the Colossus machines were dismantled and "sanitized" to remove all traces of their functioning. The secrecy around Colossus and Flowers's work, which had been involved in the decryption of the famous Enigma code, was finally broken in 1974. Between 1993 and 2007, a team of engineers led by Tony Sale used a variety of sources and declassified information to reconstruct the Mark 2, which is now on permanent display at the National Museum of Computing in Bletchley Park.

The speed of the Mark 2 is equivalent to 5.8 MHz and considerably slower than the typical speed of a modern laptop computer: 2700 MHz (2.7 GHz). But unlike modern computers, the Mark 2 did not have storage capability for programs—no RAM. Modern supercomputers like the ones that have revolutionized astronomical calculations have speeds measured in floating point operations per second (FLOPS). At this scale, the Mark 2 would achieve about eight FLOPS, while an advanced supercomputer such as the Summit computer, built in 2018 by the US Department of Energy at Oak Ridge National Laboratory, can reach two hundred thousand trillion FLOPS (two hundred petaFLOPS).

Computers are foundational to modern astronomy. At cosmic levels of measurement, precision is key. The Colossus was a huge leap forward in computing power and capability—we've come a long way since the Antikythera mechanism, the first analog computer.

49

The Radio Interferometer

A powerful breakthrough in searching the cosmos

1946

Soon after the detection of radio waves from space, radio astronomy, using a single dish antenna, underwent a revolution with the application of the interferometer method previously used by optical astronomers to measure the diameters of stars. Martin Ryle, a British astronomer, and Derek Vonberg, an electrical engineer, developed the idea of combining several radio telescopes into an array with much higher resolution. This led to the construction of the first radio interferometer at the Mullard Radio Astronomy Observatory just outside Cambridge, England.

An interferometer works by receiving radio signals from two different telescopes and combining their signals in-phase. The result is an aperture whose diameter equals the separation between the two telescopes. As with any mirror, the larger the diameter of the mirror, the more you can resolve features in the object being studied. Early generations of radio telescopes could only muster resolutions capable of showing features in the sky of roughly the angular diameter of the full Moon (0.5 degrees), and they had to be many meters across, because

radio wavelengths are the longest wavelengths in the electromagnetic spectrum—long enough that they're often measured in centimeters, and sometimes even meters or kilometers. With interferometry, you could place two radio telescopes a mile or more apart and achieve resolutions comparable to an optical photograph. During the 1950s and '60s, many such interferometers were built in England and Australia. Then in 1972, construction began on a twenty-seven-dish Very Large Array near Socorro, New Mexico, which was completed in 1980. It provided a steady stream of high-resolution images of star-forming regions, quasars, and radio galaxies at 0.2 to 0.04 seconds of arc—far higher than conventional optical photographs, and with enough resolution that many optical objects can now be uniquely identified and mapped as radio sources.

Similar interferometer systems are now available at submillimeter wavelengths, such as the Atacama Large Millimeter Array (ALMA) in the Atacama Desert of northern Chile, which is being used to study the formation of planets in dusty disks of circumstellar gas.

There is no technical limit to how large radio interferometers can be made. In the 1970s, the technique of Very Long Baseline Interferometry (VLBI) was developed using telescopes in England and the United States spanning a transatlantic baseline of thousands of miles. The radio signals were recorded on analog video tape along with an atomic clock time signal; then the tapes were brought together and replayed to correlate and phase-shift their information. The result was a radio telescope capable of resolving details as small as a millionth of an arc second. This resolution is equivalent to reading the text in this book if it were placed on the moon. Many new discoveries about the jetlike structure of quasar radio sources were made using this technique.

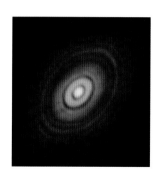

The planet-forming disk surrounding the star HL Tauri ▶

50

The Heat Shield

**Returning payloads safely
back to Earth**

1948

Over twenty thousand silica tiles cover the Shuttle Orbiter. Because of their low density, they can be easily damaged, so hundreds of these tiles must be replaced after each flight.

Sending humans up into space was undoubtedly one of the greatest achievements in the history of space exploration, but the flip side of the journey—their safe return back down to Earth—was perhaps even more technologically demanding. There's arguably no more perilous stage of a mission than reentry, when, as spacecraft make their way through the atmosphere at speeds of twenty thousand miles per hour or more, their exposed surfaces can endure temperatures as high as roughly three thousand degrees Fahrenheit, well beyond their safety and structural limits.

There are essentially two ways to deal with the heat: either an ablative heat shield or a heat sink. Heat shields are protective coverings that are meant to heat up to their melting point and then ablate, or burn off, from the spacecraft, taking their thermal energy with them. They were first used during the Bumper Program, a series of unmanned, two-stage rocket launches between 1948 and 1950. The rockets were equipped with nose cones covered with Teflon, which melted and ablated as the cone was heated to over two thousand degrees Fahrenheit at speeds of Mach 9.

The Teflon might have successfully dissipated heat, but manned flight raised the bar, requiring temperatures low enough for a human to survive. For Project Mercury, the US's first manned spaceflight program, scientists engineered a heat shield made of layers of fiberglass and aluminum, which, combined with a coolant system within the capsule, kept temperatures between eighty-five and ninety-five degrees Fahrenheit—still hot, but livable!

Similar ablative heat shields were used in the Gemini and Apollo programs. But the space shuttle *Orbiter*, with its unique aerodynamic shape, which changed the angle of reentry and therefore the frictional heat forces bombarding it, required a different solution. Here's where the second method, the heat sink, comes in. Heat sinks are simply materials that can absorb extremely high temperatures, then radiate it away as infrared radiation. The shuttle's exposed surfaces were covered in silica blocks, with carbon-fiber cloth on the leading edges of the wings—both materials can withstand the heat of reentry without melting. Instead, they radiate this heat energy very efficiently back into space. Furthermore, the materials have very low thermal conductivity, which means that their undersides, in contact with the skin of the shuttle, remain very cool. The shuttle requires twenty thousand silica tiles to keep cool.

51

The Integrated Circuit

Paving the way for spacecraft-ready computing power

1949

Before the 1950s, electronic systems such as radios and computers were powered by vacuum tubes and the heavy and bulky frames they resided in. A state-of-the-art computer capable of calculating launch trajectories in real time contained thousands of vacuum tubes, filled up large ventilated rooms, and weighed tons—a big deal for space exploration, since the biggest cost in space travel is the price per pound to place a payload into orbit. For chemical-fueled rocket technology, this number has hovered at $10,000 during much of the twentieth century. If your goal was to place humans in space, you could not fill up your launch budget with the flight computer! Fortunately the inevitable progress of electronics caught up with the advent of the space program in the late 1950s.

The first and most costly element to be replaced was the vacuum tube, made possible by the invention of the transistor by American physicists John Bardeen, Walter Brattain, and William Shockley in 1947. The transistor is a very different technology, based upon the properties of materials called *semiconductors*. They, like vacuum tubes, act like switches but consume far less electricity and remain very cold as they operate. More important, they can switch from an on (1) to an off (0) state in microseconds or less. The first transistor-based computer was TRADIC, built in 1954 by Jean Howard Felker of Bell Labs for the US Air Force. This led to very lightweight and fast computers by the time the NASA space program needed them for the Gemini and Apollo flight computers and other advanced satellite systems.

The early circuitry designs were a hodgepodge of discrete components soldered together onto wired circuit boards called a *chassis*. But a new way of fabricating electrical circuits was rapidly evolving. First came the printed circuit boards of the 1930s and '40s, in which wires were replaced by etched copper "traces" on an index card–sized plastic board. Then, in 1949, the German engineer Werner Jacobi patented an entirely new approach called the *integrated circuit*. This led to

◄ TRADIC computer

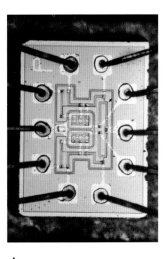

▲
Apollo Guidance Computer
integrated circuit

several refinements and innovations through the early 1950s, leading to Jean Hoerni's 1957 design of the "planar process," in which devices were assembled on a single chip of silicon through layer-by-layer deposition. With the proper lithography techniques, there was almost no limit to how small transistors, resistors, and other devices could be made, and this led to the first commercially available integrated circuits in 1961, manufactured by Fairchild Semiconductor: the 900 series of micro logic gates.

The impact that integrated circuits, or ICs, have had on both civilian and military space projects still reverberates through these programs. It is now possible to crowd billions of components onto microchips only a few inches across. This allows for faster computing times, cheaper manufacturing, and reduced weight—just the kind of progress you need to support advanced space exploration on high-cost launch vehicles!

NATIONAL BUREAU OF STANDARDS

52

◀ The first atomic clock

The
Atomic Clock

Using time to measure space

1949

Having an accurate clock has always been essential for astronomers. In fact, for centuries, navigators, religious institutions, and governments have turned to astronomers to establish the answer to this simple question: What time is it? Navigators have needed to know local time on the ship in order to establish their longitude. Religious institutions have needed to know exactly when certain events, rituals, or holidays should commence. Governments have needed to know exact times not only for precision-intensive military engagements but also to establish the civil calendar. Complex gear works driven by weights that needed daily winding, or motors that needed a constant source of electricity, had long provided the answer to this need. But the accuracy of these mechanisms varied quite a bit depending on the frictional losses among the various mechanical parts. The quartz crystal clocks built by J. Horton and W. Marrison at Bell Laboratories in 1928 achieved a two parts per million (±2 ppm) accuracy over six years of operation. This means that after six days, a clock could lose or gain as much as one second. It might not sound like much, but in the ultra-precise world of astronomy, one second off is a huge liability.

Which brings us to atomic clocks. Developed starting in 1949, atomic clocks achieve extremely precise timekeeping by a clever counting method: An atom such as cesium-133 has a quantum transition at a frequency of 9,192,631,770 cycles per second. When a microwave signal matches this frequency exactly, it excites the cesium-133 atoms so that a detector can convert the atoms in the excited state into an electrical current. This can be divided by 9,192,631,770 in another electronic system to produce a pulse exactly once every second. When the cesium atoms are carefully prepared, the clock pulses can be accurate to thirty-billionths of a second every year. In other words, such a clock will lose one second of time every thirty million years.

This skyrocketing in timekeeping accuracy is a game changer in space. Since we know how fast light travels, careful measurement of how long it takes radio waves (which move at light speed) to go from one point to another (say, from a distant star, or an orbiting satellite, back to us here on Earth) allows us to measure exact distances in space—key to mapping the universe as well as maneuvering spacecraft. Extreme time precision also allows us to gaze deeper than ever into the cosmos—it's what allowed astronomers to coordinate the simultaneous use of eight telescopes around the world, multiplying their powers and yielding a historic image of a black hole (see page 205).

And our clocks are getting even more precise: In 2013, physicist Andrew Ludlow and his team at the National Institute of Standards and Technology demonstrated an ytterbium (the element with atomic number 70) lattice atomic clock with an accuracy of two parts in one quintillion. That means the clock would lose or gain less than one second since the universe was formed fourteen billion years ago.

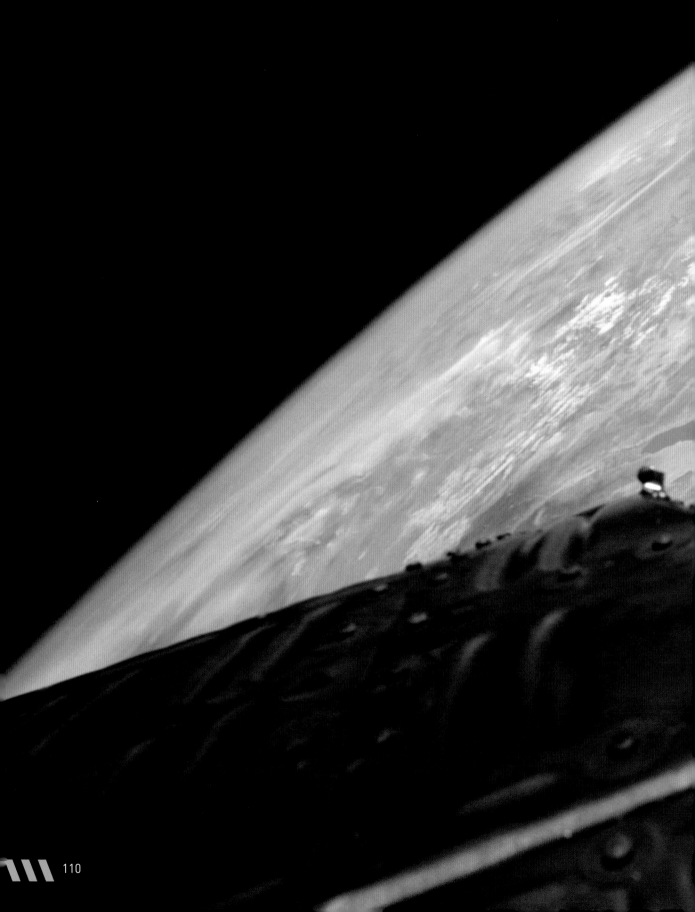

53

Space Fasteners

The unsung technology holding space exploration together

1950

The *Gemini 6* capsule is held together by numerous bolts that secure individual titanium panels.

A rocket and its payload are much more than simply a collection of panels and milled metal. No matter where you look in the design of a rocket or a scientific experiment you will encounter the unsung heroes of space exploration: the nuts and bolts that keep everything together. But of course, these are not the ordinary fasteners you might purchase at the local hardware store: They have to be space-qualified.

Under the vacuum conditions of space, and the enormous swings in temperature from near–absolute zero to over three hundred degrees Fahrenheit, metals are subject to numerous cycles of flexure, contraction, and expansion. This causes screws, bolts, and other fasteners to undergo stress and strain, which can lead to fracture and eventual failure of the components being joined. There is also the mechanical problem of vibration during launch, which can shake loose even the most securely tightened bolts. To overcome these dynamical problems, fasteners have been developed over the decades that must be far more efficient and reliable than those you that hold your furniture or car together. Much of this improvement is due to dramatic advances in aerospace metallurgy and the development of new materials that never existed before.

The most popular fasteners used in spacecraft and rocketry are made from titanium, stainless steel, or "superalloys" of nickel and chromium, such as Inconel. Each of these materials has its own specific benefits in terms of tensile strength, weight, and corrosion resistance, allowing it to be used in high-temperature rocket engines or simply to hold together an experiment package.

New generations of "SmartBolts" even have built-in strain gauges that report through color change just how much torque they are experiencing as they are being installed.

One thing is certain in the years to come: No matter where space exploration may take us, there will always be a nut along for the ride. They're the unsung heroes of space—until now!

The *Gemini 6* capsule

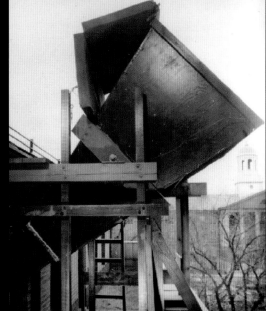

54

The Hydrogen Line Radio Telescope

Mapping the interstellar medium

1951

In 1904, astronomer Johannes Hartmann discovered that the spectrum of the star Delta Orionis had an absorption line from the element calcium (in other words, it absorbed light at a wavelength unique to calcium), and correctly interpreted this as evidence that there were clouds of gas in space that contained calcium, among other elements. By the 1940s, the development of radio astronomy aligned with Hendrik van de Hulst's theory that some of the radio noise in the sky could be coming from interstellar hydrogen gas emitting at a frequency of 1420 MHz, which corresponds to a wavelength of twenty-one centimeters. The first successful detection of this twenty-one-centimeter signal occurred on March 25, 1951, when Harvard physicists Harold Ewen

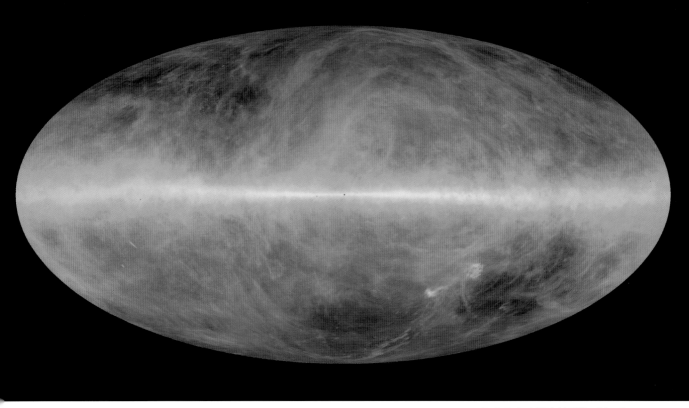

▲

This all-sky hydrogen map was produced using data from the one-hundred-meter Effelsberg radio telescope in Germany and the sixty-four-meter CSIRO radio telescope in Australia.

and Edward Purcell built an antenna and a receiver and literally pointed them out the window of their laboratory.

This so-called radio telescope was simply a horn-shaped collector, familiar to radar technology, that was pointed up at the sky to collect the weak radiation at a wavelength of twenty-one centimeters. The arrangement worked far better than Ewen and Purcell had hoped, and they were treated to a strong signal that followed the passage of the Milky Way across the sky and through the view of their horn antenna. Still, there were many technical problems that had to be overcome. When it rained, the horn filled up with water and had to be drained. (Students also enjoyed tossing snowballs into it during the winter!) Also, to avoid the enormous background noise that occurred at these frequencies, Ewen devised a frequency-switching technique in which measurements of the sky at nearby frequencies were subtracted from the signal at twenty-one centimeters, making it far easier to see this weak signal.

The first maps of hydrogen in the Milky Way were completed by van de Hulst, C. A. Muller, and Jan Oort in 1954, and by C. A. Muller and Gart Westerhout in 1957. These maps, and others like them at still-higher resolution, were used to map out the structure of the Milky Way galaxy, revealing its spiral arms and complex patterns of giant interstellar hydrogen clouds within which new stars were being formed. Modern applications of hydrogen-line astronomy are now probing the Dark Ages of cosmic evolution, in search of the first generation of stars to have formed in the universe, when it was only a hundred million years old.

55

The X-Ray Imaging Telescope

A new window onto the universe

1952

The mirrors of the Chandra X-ray Observatory ▶

It is a relatively easy thing to focus visible light using silvered mirrors, but at other wavelengths this method becomes completely unworkable, especially for short-wavelength X-rays—not an inconsequential thing, given that the universe is filled with celestial bodies sending out copious amounts of electromagnetic radiation at these shorter wavelengths. X-rays will be absorbed if they strike a surface head-on, but if they strike the surface at a glancing or grazing angle of less than two degrees, they can be made to be reflected. In 1952, Hans Wolter developed three types of grazing-incidence optical systems, which are now the mainstays of today's advanced imaging systems that work with X-ray energies.

The *Copernicus* (OAO-3) X-ray satellite, launched in 1972 as a collaboration between NASA and the United Kingdom, was the first to use a grazing-incidence system in space. The X-ray detector was built by University College London's Mullard Space Science Laboratory for a Stellar X-rays experiment conducted by Sir Robert Boyd. The two X-ray telescopes each had a collecting area of less than two square inches, and the X-ray photon counter sat at the focus of the mirrors. They were able to detect X-rays from stars and other known sources between wavelengths of one to seventy Angstroms. The data from this system spanned over eight years,

allowing the X-ray variability of some sources to be studied in detail.

In 1978, NASA launched the Einstein Observatory (HEAO-B) with four nested Wolter grazing-incidence mirrors, which allowed it to achieve optical observatory-quality imaging of astronomical objects at arc-second resolution. Instead of seeing just a blurry spot of X-rays in the sky, the Einstein Observatory could resolve fine details and detect individual "point sources" by the thousands.

X-ray optics and Wolter's designs continue to be applied in all major imaging X-ray systems today, including the Chandra X-ray Observatory, launched in 1999, and NuSTAR (Nuclear Spectroscopic Telescope Array), launched in 2012. This technology, developed all the way back in 1952, has led to spectacular discoveries in black hole physics, dark matter investigations, and the exploration of the high-energy universe—the cutting edge of space exploration today.

◀ The OAO-3 X-ray mirror

56

The Hydrogen Bomb

A destructive demonstration of how stars shine

1952

I n his 1920 paper "The Internal Constitution of the Stars," Sir Arthur Stanley Eddington proposed that the fusion of protons in the core of the Sun and other stars would produce enough sustained energy to keep stars stable against gravitational collapse. The source of this energy would be the first demonstration of Albert Einstein's famous formula $E=mc^2$. The internal temperature of the Sun would have to be tens of millions of degrees Fahrenheit in order for the protons to have enough energy to overcome their mutual electrostatic repulsion.

The fusion of four protons to make one helium nucleus would leave enough mass converted into energy to supply the solar luminosity, so it seemed as though the latent energy of fusion was there; it only

needed to be liberated. The mathematics seemed straightforward. A proton has a mass of 1.673×10^{-24} grams, and a neutron's mass is 1.675×10^{-24} grams. If you add two neutrons and two protons you get a total mass of 6.696×10^{-24} grams. An actual helium nucleus has a mass of 6.646×10^{-24} grams. The difference of 0.05×10^{-24} grams represents the binding energy of a helium nucleus, and from $E=mc^2$ is a source of energy equal to 0.000045 ergs per fusion. To light up the Sun, about four million tons of mass must be converted into energy every second.

But there was one significant problem: A helium nucleus consists of two protons and two neutrons. How do you get two of the four protons to become neutrons? The answer came from physicist Hans Bethe, who showed that a proton could be transformed into a neutron. Not only that, but the reactions could occur at a much lower temperature than the Sun's core—only twenty-seven million degrees Fahrenheit, thanks to a quantum mechanical process called *tunneling*. As a result, hydrogen would "burn" through a cycle of steps termed *the proton–proton cycle*. This cycle is the main source of energy for our Sun and stars of similar mass.

Although the proton–proton cycle applies only to astronomical bodies, the detonation of the first hydrogen fusion bomb in the Marshall Islands in 1952 showed the devastating power of only a few grams of hydrogen converted into energy. Eventually, thermonuclear fusion was investigated as a potential source of clean energy, and many scientific groups funded by government energy contracts have attempted to create controlled fusion.

We're all aware of the ways in which technology developed for use in space has, over time, trickled into our daily lives here on Earth. The hydrogen bomb represents a variation on that process, a translation of a natural "technology" already found in space, in our Sun, into a form used on Earth. It's a testament to the awe-inspiring—and profoundly dangerous—ways we apply what we learn from the study of colossal powers and forces in the universe.

57

The Radioisotope Thermoelectric Generator

Electricity when the sun doesn't shine

1954

For journeys beyond the orbit of Jupiter, sunlight is so dim that solar cells do not perform well. Luckily, long before NASA's first spacecraft, *Pioneer 1*, ventured into space in 1958, there was already a solution to this problem. In 1954, Ken Jordan and John Birden at the Atomic Energy Commission's Mound Laboratory in Ohio combined a thermocouple with a sample of radioactive polonium-210 and created the first working radioisotope thermoelectric generator (RTG). A thermocouple is a pair of dissimilar metals joined together that create electrical currents at their fused junction when heated; the polonium-210 isotope generated heat through the decay of its atoms. When these parts were combined, electricity could be generated at a rate of 140 watts per gram, but the half-life of polonium-210 was only 138 days, so after that period of time, the power was reduced by a factor of two.

The first use of RTGs in space was on the US Navy's Transit 4A satellite, successfully launched in 1961. Its 2.7 watts of power were modest, but it was able to avoid having to use large solar cell arrays. It was a promising start to the use of nuclear energy for peaceful purposes, but in 1964, a satellite powered by an RTG failed to reach orbit, and two pounds of plutonium-238 fuel disintegrated in the atmosphere over the southern hemisphere. A decade later, John Gofman of the University of California, Berkeley, claimed this enhancement in atmospheric plutonium may have actually increased worldwide lung cancer rates. This incident caused NASA to develop its solar panel technology in earnest rather than rely solely on plutonium-based RTGs for its satellites—a choice not made by the Soviets in their RORSAT-series satellites.

Nevertheless, RTGs were invaluable for the Apollo program lunar experiments, as well as occasional missions to Mars and the outer solar system. The most successful model for NASA was the SNAP-19, which was used on the *Pioneer 10* and *11* and *Viking 1* and *2* missions of the 1970s. And the SNAP-27 was used by the Apollo mission to power its lunar surface science experiments. Because of the lack of sunlight, RTGs are used on all deep space missions such as Galileo, Cassini, *Voyager 1* and *2*, Ulysses, and New Horizons. Most RTGs use plutonium-238 with a half-life of eighty-eight years. The Voyager spacecraft, now more than forty years old, have lost half their power, but still produce enough electricity to keep a few instruments operating from beyond the orbit of Pluto.

Artist's rendering of the Transit 4A satellite in orbit ▶

58

The Nuclear Rocket Engine

Now we're getting somewhere!

1955

While the use of nuclear energy to power a spacecraft's internal functions has found its place in space, using nuclear power in far higher-energy rockets has taken a different path. Rockets are actually very simple devices. All they do is throw as much mass as possible out the back of the engine nozzle so that the spacecraft will be pushed to high speeds in the opposite direction. It's all about how much mass you can emit in a given time. For thousands of years, controlled chemical combustion (gunpowder, liquid fuels) was the only way to create high-speed flows of large amounts of matter. It boils down to exhaust speed and exhaust mass: The combined product is what gives a payload the momentum and high velocity needed to break away from a planet's gravity or to maneuver in space. But is there any other way to impart a lot of speed to matter to make a rocket go? In the post-nuclear world of the 1950s, the answer was an emphatic yes!

The first nuclear rocket engines were tested at Los Alamos Scientific Laboratory (as LANL was then known) as part of Project Rover, run by NASA and the Atomic Energy Commission between 1955 and 1972. The Kiwi-B nuclear thermal engine was powered up in December 1961. The fuel was simply liquid hydrogen passed through a small nuclear reactor and heated to 3,700 degrees Fahrenheit. It created 1,100 megawatts of heat energy and a thrust of 25 tons (50,000 pounds). Compared to a typical chemical rocket capable of over 750 tons of thrust, nuclear engines were not a good bet for getting off a planet, but in low-gravity space they came into their own.

Testing of nuclear rocket engines continued until 1968, when the last and most powerful nuclear engine, the Phoebus-2A, was powered to full throttle for twelve minutes, achieving an impressive 930,000 Newtons (210,000 pounds) of thrust. A year later, NASA's Wernher von Braun designed a nuclear-powered expedition to Mars to be launched after the end of the Apollo program, but in January 1973, all funding for the nuclear rocket program, now known as NERVA, was terminated, so von Braun's plan never came to fruition. The nuclear rocket program was deemed a success, but national priorities changed dramatically after the cancellation of the Apollo program, in favor of investment in a space shuttle and an orbiting space station.

◀ Phoebus-2A reactor
in operation

59

Sputnik

Russians win the space race
. . . for a few months

1957

A replica of *Sputnik 1*, the first artificial satellite to reach outer space, at the National Air and Space Museum ▶

The events of October 4, 1957, shocked the world. Seemingly out of nowhere, the former Soviet Union launched *Sputnik 1*, a 184-pound, 23-inch-diameter satellite, into orbit. Every ninety-eight minutes, the signal from its one-watt transmitter could be heard near frequencies of 20 and 40 megacycles (called *megahertz* today) by any radio transmitter on Earth. Operation Moonwatch had over 150 stations manned by amateur astronomers watching for the faint starlike satellite as it raced across the twilight sky. Amateur radio operators of the American Radio Relay League, meanwhile, listened to their radio receivers for the distinctive *beep-beep-beep*. From the vantage point of today, we imagine that this one event triggered the massive space race between the United States and the USSR, but in fact President Dwight D. Eisenhower, although knowledgeable about the USSR's progress via U-2 reconnaissance planes, was by some accounts dismissive of the event. Even the USSR did not use it initially in its propaganda. Eisenhower greatly underestimated the reaction of the American public, however, who were shocked by the launch of *Sputnik 1* and by the televised failure

of the US's Vanguard Test Vehicle 3 launch on December 6, 1957. Only the successful *Explorer 1* spacecraft, launched on January 31, 1958, got the United States back in the game.

The *Sputnik 1* satellite was quickly followed by much better funded progress in the space race, having been the catalyst for the creation of both NASA and the Advanced Research Projects Agency (later renamed the Defense Advanced Research Projects Agency) in 1958. There was also huge increases in funding for education and scientific research.

Sputnik 1 lasted twenty-two days before its battery power ran out, and it went on to orbit for seventy-one more days before it burned up in the atmosphere on January 4, 1958. During its short life, the upper-atmosphere drag on the satellite could be measured by noting the satellite's orbital speed change at perigee. This gave scientists valuable information for the first time about the upper atmosphere's density and its variation with altitude. Also, by studying the frequency shift and propagation of the radio signals at 20 and 40 megacycles, the properties of the topside ionosphere (an ionized outer layer of the atmosphere) could be measured from space.

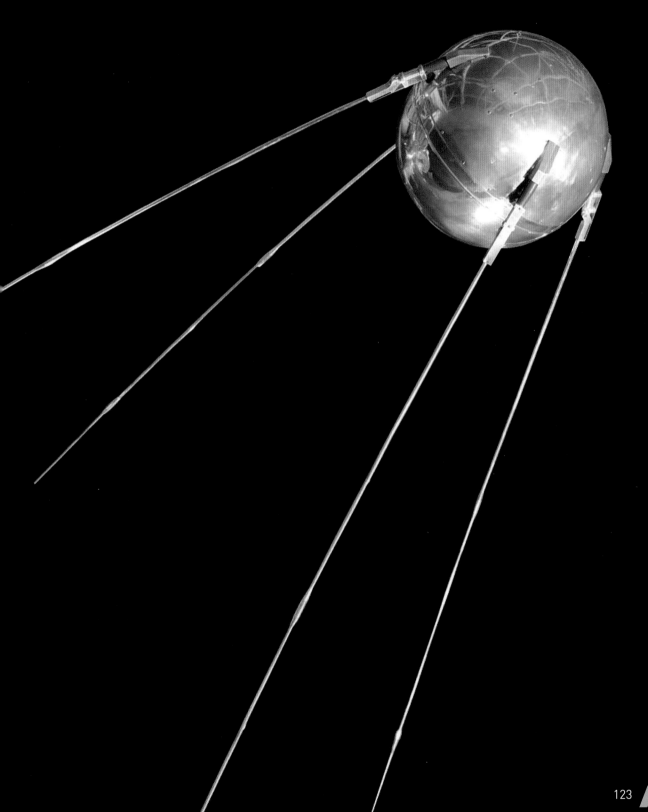

60

Vanguard 1

The oldest space junk

1958

On March 17, 1958, *Vanguard 1* became the fourth artificial satellite successfully launched into orbit. Manufactured by scientists and engineers at the US Naval Research Laboratory in Washington, DC, this grapefruit-sized satellite was the first to use solar cells to generate electricity. The early years of satellite launching were highly experimental: Vanguard represented a pioneering design for satellites, and it came at a time when there was still considerable debate over a satellite's optimal shape. The first successful US satellite, *Explorer 1,* was cylindrical. Some designers favored a conical shape, while others argued that a uniform sphere was the best shape for providing uniform drag through the atmosphere.

Vanguard 1 was arguably most significant as a political tool. The Soviet Union had launched *Sputnik 1* on October 4, 1957, only five months earlier, which took Americans completely by surprise and marked the beginning of the so-called space race. The United States answered the challenge by successfully launching *Explorer 1* in January 1958, followed by *Vanguard 1* a little more than a month later. Between *Sputnik 1* and *Vanguard 1,* several US satellite launches failed (*Vanguard 1A, 1B,* and *Explorer 2*), which was not only scientifically devastating

but politically embarrassing. After all, in the eyes of the agencies that launched them, these endeavors were yardsticks of the relative might of communism and democracy.

Politics aside, *Vanguard 1* was noteworthy for its engineering and scientific accomplishments. The world's first solar-powered satellite, its six solar cells provided power to its very modest, five-milliwatt transmitter, which continued to operate for six years, sending data back to the ground about the electrons and radiation it encountered high above Earth's atmosphere. Scientists also used its orbit to discover that Earth is slightly pear-shaped, being a bit narrower near the North Pole and a bit fatter near the South Pole.

But perhaps the satellite's most lasting mark on the history of space exploration was due to its highly elliptical orbit, swooping down to 400 miles at perigee but reaching close to 2,500 miles at its apogee point, taking over two hours to complete a full circuit. This ensured that *Vanguard 1* would experience very little atmospheric drag, and that it would remain in space for more than two hundred years. In fact, by 2018 it had become the longest-surviving artificial object in space—the oldest of the thousands of objects that now litter the heavens.

But *Vanguard 1* is more than just another piece of space trash. Thanks to its staying power, for more than fifty years scientists have been able to study the satellite's minute orbit changes as it interacts with Earth's outer atmosphere. From this they've developed a better understanding of how Earth's atmosphere changes its shape and density over time, which has led to insight into such issues as satellite signal transmission and even climate change.

Sometimes even junk, when properly used, can reveal the secrets of the universe!

The debris cloud of junk surrounding Earth

▼

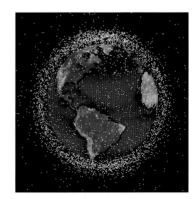

61

Luna 3

Our first glimpse of the Moon's far side

1959

The lunar far side as seen by *Luna 3* ▶

Launched on October 4, 1959, by the former Soviet Union, this small spacecraft—just over four feet long and weighing 613 pounds (278 kilograms)—followed its predecessor, *Luna 2*, to the Moon. *Luna 2*, launched a few weeks earlier, has its own entry in the history of space as the first human-made object ever to land on a celestial body. But *Luna 3* was arguably the more historic. Its mission was simple: During its flyby of the Moon, it was to take as many pictures of the far side as possible.

Far side, back side—however it is that we refer to the side of the Moon that always faces away from us, we should remember that another popular name for that side, the "dark" side, is a misnomer; it's lit by the Sun for just as long as the side we can see. The Soviets timed *Luna 3*'s launch to take advantage of this; while the Moon appeared entirely dark from Earth—in its new moon phase—the far side was fully illuminated by the Sun. Following a two-day voyage of 40,500 miles (65,200 kilometers), the spacecraft detected the Moon's light and automatically opened its camera lens shutter. Over a period of forty minutes on October 7, *Luna 3* took twenty-nine photographs using its conventional film camera system.

But not all of them were successfully transmitted back to Earth. *Luna 3* had an onboard photo development system and a primitive scanner to send something like faxes of the photos back home. This resulted in photos of varying quality, and twelve were never received; only six were ever published.

But that was enough to make history: These grainy images were humankind's first-ever glimpse of the far side of the Moon. As crude as the photos were, they revealed something surprising that had astronomers scratching their heads for the next five decades: Unlike the near side of the Moon, with its large, dark plains—the remains of volcanic eruptions known as *lunar maria*—and bright, cratered highland regions, the back side of the Moon had no such maria. Only a handful of very small dark splotches could be discerned, the largest of which were named Mare Moscoviense and Mare Desiderii. Whatever mechanism caused the enormous mare of the lunar landscape on the side facing Earth, it was, mysteriously, largely absent from the far-side surface.

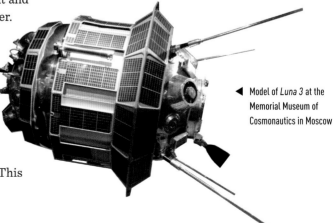

◀ Model of *Luna 3* at the Memorial Museum of Cosmonautics in Moscow

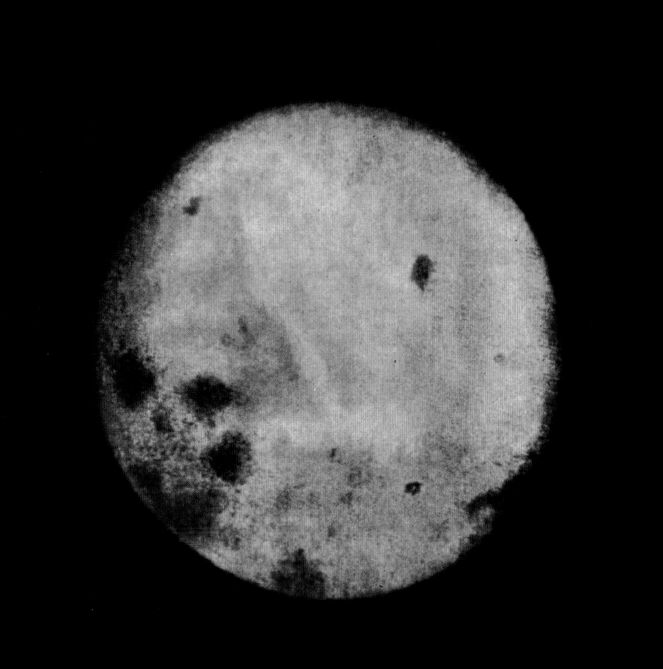

62

The Endless Loop Magnetic Tape Recorder

Data storage in space

1959

An engineering prototype of the TIROS-1 magnetic tape data recorder, ca. 1960, displayed at the National Air and Space Museum ▶

◀ The *Vanguard 2* satellite, magnetic tape recorder within

The tape recorder didn't significantly enter the general public consciousness until Bing Crosby took a deep interest in the technology, spurring a huge uptick in consumer sales. But the technology for magnetic tape recording had been around long before his time. The first non-magnetic tape recorder, invented by Alexander Graham Bell, along with his cousin Chichester and technician Charles Sumner Tainter in 1886 (the latter two of whom are listed on the patent, No. 341,214), consisted of a $^3/_{16}$-inch paper tape covered with beeswax and paraffin passing under a moving stylus. About a decade later, Valdemar Poulsen developed magnetic sound recording by imprinting the changes in a magnetic field onto a wire. This was followed by the photoelectric paper tape recorder of 1932, in which electrodes embedded stripes representing soundwaves directly onto chemically reactive paper. Then came the Magnetophon, developed by the German company AEG in partnership with BASF, which made the lightweight magnetic tape using a paper coated with magnetically responsive iron oxide powder to record sound waves; this allowed the machine to be smaller and less expensive. Presented to the world at the 1935 Radio Fair in Berlin, it was the world's first practical tape recorder.

But why would this technology have anything to do with the exploration of space? Because data storage dramatically increased the amount of information we could gather from a single spacecraft.

Satellites were designed to capture data in a variety of modes, including photographs and simple analog data from sensors. The problem was that in the early years of space exploration, data were produced faster than the telemetry system could transmit the signals to the ground. Furthermore, few ground stations existed to receive the data, so scientists had to find some way to store information between telemetry downloads. They found their answer in magnetic tape, the only technology available for data storage in the late 1950s and early '60s.

The *Vanguard 2* satellite, launched on February 17, 1959, was the first satellite to use magnetic tape technology. Its successful launch was considered a major accomplishment for the United States in the space race because it was an answer to the Soviet Union's *Sputnik*. The goal of *Vanguard 2* was to measure the amount of cloud cover in the daytime for a period of nineteen days, after which its battery would run out. It used a one-watt transmitter at 108 MHz and a tape recorder (one that uses tape that records continuously, without rewinding) to record fifty minutes of data and play it back in a one-minute burst.

Magnetic tape recorders filled a crucial niche in spacecraft data storage and buffering (storage while data is being processed or transferred) and directly contributed to the scientific successes of the TIROS and Nimbus weather satellites, as well as the Mariner, Viking, Galileo, and Voyager spacecraft programs.

63

The Laser

A new light and way of seeing

1960

Components of Maiman's original laser ▶

◀ Laser guide star system on the Very Large Telescope of the Paranal Observatory

Imagine you're at a party on the second floor of an apartment building, but the only access is by elevator to the third floor; from there, you have to walk down a flight of stairs. People keep arriving over the course of an hour, but then the landlord complains and shuts the party down. Everyone leaves by taking the stairs to the ground floor in a rush. This is the basic principle at work in the process of light amplification by stimulated emission of radiation (LASER), a phenomenon that was well-known by 1960 but hadn't yet been configured into a practical, easy-to-use device.

A laser works by pumping electrons into an excited state. That state decays rapidly, but the electrons are then held up for a long time in an intermediate "metastable" state before making their last decay to the ground state and releasing a final photon of light. Because the second photon has a unique and precise wavelength, it's visible as a consistent, coherent beam of light.

As Hughes Research Laboratories physicist Theodore Maiman discovered, a polished cylinder of ruby stimulated by a high-intensity flash lamp was enough to make ruby "lase." Many researchers were working on a continuous pumping process, but Maiman realized that a brief flash of light was enough to adequately stimulate the electrons when done rapidly and frequently. The first optical laser is credited to his perfection of this device on May 16, 1960.

He announced it to the world in a press conference on July 7 of that year.

Since then, lasers have been incorporated into numerous applications, from printers and industrial metal-cutting systems to high-precision optical metrology systems. Space exploration relies on lasers in a growing number of ways. The first interplanetary laser communication system was tested in 2001 with the European Space Agency's *Artemis* satellite. In 2005, the laser altimeter on NASA's *Messenger* spacecraft communicated with Earth across fifteen million miles. The OPALS (Optical Payload for Lasercomm Science) experiment on the International Space Station achieved a laser transmission speed of fifty megabits per second in 2014. Lasers are used for the precision fabrication of parts for spacecraft; they're also used by the *Curiosity* rover to vaporize rocks for chemical analysis.

Perhaps the most exciting application of lasers in astronomical research is to provide artificial stars for adaptive optics. A ground-based telescope equipped with adaptive optics can completely cancel the twinkling effects of Earth's atmosphere. To do this, the system uses lasers to create bright spots of light in the sky whose properties are precisely known. The twinkling disturbances in the laser guide stars can then be used to cancel the disturbances in the images from distant stars and planets.

A Skylab food tray. Out of tray, from left to right: sugar cookie cubes, beef sandwiches, chicken and rice, beef pot roast, and grape drink. In tray, clockwise from top right: strawberries, asparagus, prime rib, dinner roll, butterscotch pudding, orange drink.

64

Space Food

Cuisine for the Space Age

1961

In 1961, the Soviet cosmonaut Yuri Gagarin ate from three toothpaste-type tubes, two filled with servings of puréed meat and one with chocolate sauce. Since then, space food has steadily evolved to be more palatable, lightweight, and nutritious while dealing with the unique challenges of a microgravity environment. Fluids do not pour, and liquids tend to form into globules, while crumbs can become dangerous sources of pollution that might interfere with electrical systems. At first, most meals were extruded from tubes, but over the years of operating the International Space Station (ISS), small kitchens have been installed with convection food warmers for heating food and the equivalent of sinks for obtaining hot water and rehydrating dried foods.

▲
Italian ESA astronaut Samantha Cristoferetti preparing espresso in space

An entire scientific discipline has evolved to encompass the psychological and biological needs for food and food preparation in confined, micro-g environments. Food in space isn't merely about the fulfillment of caloric needs; it is also an important part of the social and psychological life of ISS inhabitants during long missions. Some foods are banned (because of their potentially noxious odors, for instance), while others are favored: Astronauts have a preference for spicy foods because their sense of taste tends to diminish in space. On the Skylab space station in the 1970s, shrimp cocktail and butter cookies were consistent favorites. Lobster Newburg, fresh bread, processed meat products, and ice cream were among other choices. On the ISS, the many different canned foods and fresh vegetables have been supplemented by specific requests from the international crew. Recently, a modified version of Korea's national dish, kimchi, made it to space. It took three research institutes several years and millions of dollars in funding to create a version of the fermented cabbage dish that was suitable for space travel. The Russian crew has a selection of over three hundred dishes to choose from. In 2007, Swedish astronaut Christer Fuglesang was not allowed to bring reindeer jerky with him on a space shuttle mission. The American astronauts thought it was "weird" so soon before Christmas, so he had to go with moose jerky instead.

One of the most recent advances in space cuisine takes the form of a coffee maker called the ISSpresso, developed by Argotec. Reflecting on the first batch of espresso served on the ISS in 2015, astronaut Samantha Cristoforetti tweeted that it was "the finest organic suspension ever devised. Fresh espresso in the new Zero-G cup! To boldly brew. . . ." Given that sunrise on the ISS happens every ninety minutes, NASA does not recommend a morning cup of espresso to start the day!

65

The Space Suit

A life-supporting second skin

1962

◀ John Glenn wearing a
Mark IV space suit

The Soviet cosmonaut Yuri Gagarin made his historic orbital flight on April 12, 1961, becoming the first person in space. The United States responded to that geopolitical event with an urgent mission of its own, sending Alan Shepard into suborbital spaceflight on May 5, 1961, and then, finally, John Glenn into full orbital flight on February 20, 1962.

Unlike the roomy *Vostok 1* capsule used by Gagarin, Glenn's *Friendship 7* capsule was cramped. In case he landed in the South Pacific upon reentry, he carried a note written in several languages that read "I am a stranger. I come in peace. Take me to your leader and there will be a massive reward for you in eternity." The following year, the Soviets launched the first female space traveler, Valentina Tereshkova. This historic event would not be answered by NASA until astrophysicist Sally Ride flew on space shuttle mission STS-7 on June 18, 1983.

The spacecraft technology in these missions varied widely. But one common denominator, without which none of these astronauts would have reached space and lived to tell about it, was a functioning space suit.

Since the invention of the pressure suit in the 1930s, the US Air Force had used many kinds of flight suits for its pilots who flew above the Armstrong limit of sixty-two thousand feet, where water and other fluids begin to boil at body temperature due to low pressure. By the late 1950s, the US Navy's Mark IV pressure suit, worn by jet pilots

during the Korean War, became the favored suit because it wasn't especially bulky or heavy, and most important, it was much more mobile. The Mark IV was then adopted and adapted by NASA for Project Mercury (1958–63); it was used in John Glenn's historic flight. Subsequent space-suit designs included wrist bearings and an all-important urine collection system.

Because a pressurized suit tends to balloon outward in space, straps had to be added to prevent the expansion of the suit. Also, moving one's wrists, neck, and elbow joints is difficult when pressurized; rotating bearings eventually had to be added so that astronauts could actually move and do work. The suits were made from silver or white material to reflect sunlight; darker suits tended to heat up so quickly that they were uninhabitable after a few minutes. By the time the Gemini program launched in the 1960s, a cooling system had been added that circulated a fluid throughout the suit via an internal garment. This increased the bulkiness of space suits used for extravehicular and lunar excursion activity. Ordinary flight suits worn for emergency conditions during launch and reentry still resembled the slim and lightweight Mercury suit designs.

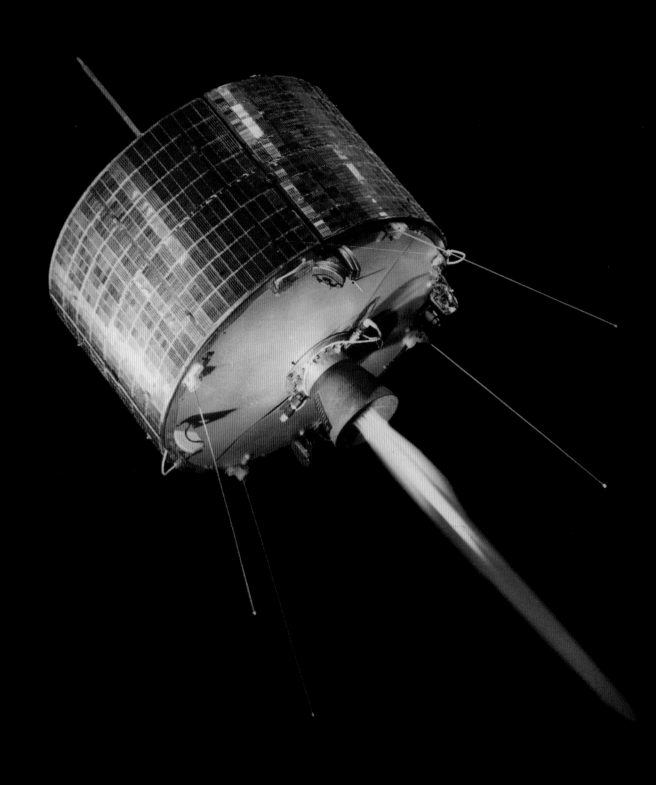

◄ *Syncom 2* satellite

66

Syncom 2 (and 3)

Making space commercially accessible

1963

A geosynchronous orbit is one that matches the speed of Earth's rotation. So a geosynchronous satellite would remain in one place overhead, night and day, relative to the surface of Earth. The Hughes Aircraft Company set about capitalizing on the potential for communications with such orbits, in which a satellite could constantly beam information down to the same place, with its Syncom program in the 1960s.

The first Syncom satellite failed before reaching its final orbit, but the second, *Syncom 2,* launched on July 26, 1963, became the world's first geosynchronous communications satellite. NASA ran voice, fax, and teletype tests and then made history again by sending the first geosynchronous-originated TV transmission as well as the first satellite-linked communication between heads of state, when President John F. Kennedy called the prime minister of Nigeria a month later. The way information was sent around the world would never be the same.

Syncom 3 deserves recognition, too, for taking the technology much further and popularizing the power of geosynchronous communication with a mass audience. Launched on August 19, 1964, *Syncom 3* was a cylindrical satellite measuring only fifteen inches tall and twenty-eight inches in diameter. It was covered with 3,800 silicon solar cells to generate the twenty-nine watts needed to operate its two transponders. It set its own record as the first truly geostationary satellite—a satellite in orbit both geosynchronously *and* directly over the equator—having reached a distance from the center of Earth of 26,199 miles over the equator at a longitude of 180 degrees east over the Pacific Ocean.

The satellite was used for a variety of tests, including televising the 1964 Olympic Games in Japan and relaying teletype transmissions for airlines on the San Francisco-to-Honolulu route. In January 1965, ownership was turned over to the Department of Defense, which used *Syncom 3* during the early years of the Vietnam War.

Although Canada, Europe, Japan, and the United States collectively paid $1 million for access to the satellite during the Olympic Games, Canada and Europe saw more of the broadcasts transmitted by *Syncom 3,* because NBC preferred to use videotapes flown in from Tokyo every day to provide a higher-quality television experience. But any American willing to stay up until 1:00 AM eastern time on the night of the opening ceremony could see it live from Japan.

The satellite had to be made very lightweight in order to reach orbit, so there was no audio on the *Syncom 3* rebroadcast. Instead, the audio content was sent via transpacific cable; the audio and video streams were received at a ground station in Point Mugu, California, then relayed to suburban Burbank, where they were combined and synced.

67

The Vidicon Camera

Electronically imaging objects in space

1964

One of the most basic astronomical activities is photography. But in the space age it is one thing to photograph what a telescope is seeing and quite another thing to relay that information via radio signals millions of miles back to Earth. The earliest attempt at image telemetry was 1959's *Luna 3* moon-imaging spacecraft, which involved taking a photograph with a film camera. Once the negative was chemically processed, it would be electronically scanned and the signals transmitted back to Earth. But there was another way to do this that didn't involve chemicals or film: television!

Many ground-based attempts at transmitting images electronically were made before the 1920s, but it wasn't until 1926 that the Hungarian engineer Kálmán Tihanyi patented a method that actually worked. The basic idea

◀ Mariner Crater—diameter of ninety-four miles—near Phaethontis quadrangle (note the pixelization)

The Voyager vidicon tube—gift of the NASA Jet Propulsion Laboratory, California—currently at the Herschel Museum of Astronomy, Bath, UK ▼

was that an image would be focused on the surface of a specially designed vacuum tube with a flat surface coated with a light-sensitive material like selenium. When light fell on the surface, electrons would be generated in numbers proportional to the light intensity. Then a cathode ray would scan this surface to "read out" the image. RCA developed image orthicon tubes in 1946, which were steadily perfected so that they worked with progressively lower light levels, leading to a new design called the *image vidicon tube*. For spacecraft operations, a telescope would focus an image on the vidicon tube, and the image would be electronically scanned and telemetered to Earth without needing any intermediary film-processing steps.

The Television Infrared Observation Satellite (TIROS-1), launched in 1960 as a weather satellite, proved that this new vidicon technology actually worked in space. *TIROS-1* was followed by the *Ranger 3* spacecraft in 1962, which was equipped with a video system but failed to reach its target, the Moon. But one of the most famous early successes of vidicon imaging technology took place during the *Mariner 4* mission to Mars, launched in 1964. The surface of Mars was the subject of much speculation at the time, whether you were a child or a professional astronomer, and it was fully expected that *Mariner 4*'s imagery would at last answer this fundamental question: Were there really canals on Mars? The answer that came back from the twenty-two pictures and 634 kilobytes of data that were telemetered to Earth

was not what anyone had expected: There were no canals. Just craters.

For *Mariner 4*, the video camera developed by the Jet Propulsion Laboratory was mounted at the focus of a small 1.5-inch telescope with a one-degree field of view and a resolution of about two miles. The telescope would image the surface of Mars onto a vidicon tube, which would convert the light intensity changes into an electrical signal. The intensity of this signal was then digitized into six bits (sixty-four intensity levels) and 240,000 bits per 200-by-200-pixel picture. The telemetry to Earth was slow at such a great distance, so an onboard magnetic tape recorder with a 330-foot tape loop and a five-million-bit capacity stored the images until they could be transmitted at about eight bits per second.

The imaging vidicon technology allowed many space missions to make their historic discoveries, such as *Pioneer 10* and *11, Viking 1* and *2,* and *Voyager 1* and *2*. The last time a vidicon system was used in this way was for the Voyager spacecraft, which was launched in 1977, but designed and fabricated starting in early 1972 as the Mariner Jupiter/Saturn Project. By 1975, digital camera technology had evolved to the point that NASA decided to use a solid-state imaging camera on the Galileo mission, launched in 1989; digital imaging was also integral to the Hubble Space Telescope's 800-by-800-pixel wide field/planetary camera, which was selected for Hubble's imager in 1982 and finally launched in 1990.

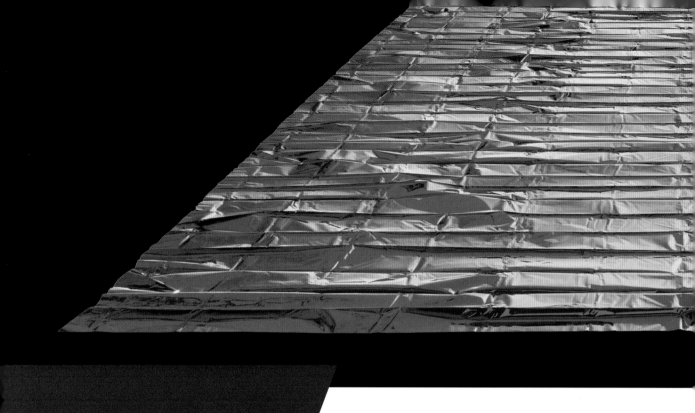

68

The Space Blanket

A simple way to keep heat in—and out

1964

Some space technologies not only further the advancement of exploration but also dramatically improve the lives of earthbound citizens. One of the lowliest and most unsung technologies is a simple piece of plastic coated in a reflective metallic film. Engineers use these so-called space blankets as part of spacecraft thermal regulation.

Thermal blankets were developed in 1964 at the dawn of the US space program by engineers at the NASA Marshall Space Flight Center. Creating these metallized films is not a trivial process: The vaporized aluminum has to be matched to the properties of the mylar plastic so that infrared waves (heat) can be redirected rather than conducted. When used with the reflective film facing the body, up to 97 percent of the infrared energy falling onto the body can be reflected back to keep the body warm. When reversed, with the reflective film facing outward, the blanket can also be used as a mirror-like reflector to keep the human body cool by reflecting the infrared energy away.

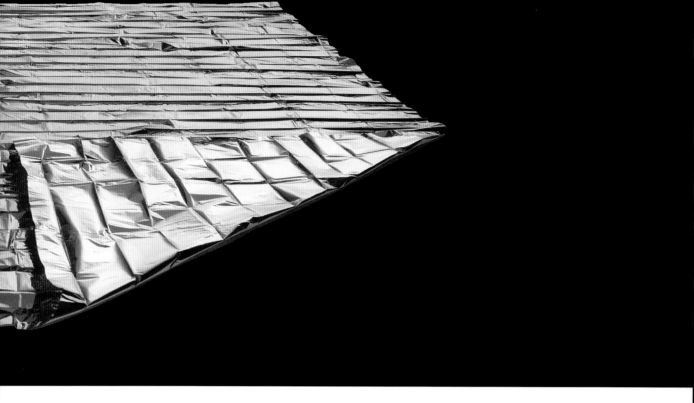

Space blankets aren't just used to regulate body heat; in fact, nearly all spacecraft, manned and unmanned, use them—from the gold reflective blankets wrapped around the base of the Apollo lunar landers to the Hubble Space Telescope and the Mars rovers. Perhaps the most famous example of the crucial role that thermal blankets play in space came during the Skylab mission in 1973, when an external sunshade failed. Astronauts had to jury-rig a new one made from a space blanket in order to reduce the interior temperature to survivable levels.

These days, space blankets have become a must-have safety item for campers, explorers, and other recreational adventurers. At the end of 1979's New York City Marathon, runners were given these space blankets to avoid hypothermia. They are now handed out at finish lines everywhere.

A sunshield used during the *Skylab-3* mission ▶

69

The Handheld Maneuvering Unit

Getting around in space

1965

Handheld maneuvering unit ▶

Ed White's handheld self-maneuvering unit used during extravehicular activity (EVA) on *Gemini 4* flight ▶

It was a sunny day in June 1965 when astronaut Edward White decided to exit the cramped *Gemini 4* capsule and go out for a walk—in space. While his partner, astronaut James McDivitt, stayed inside the capsule, strapped into his seat in a space suit, White exited with his life-supporting umbilical cable (which provided oxygen and a communication link) in tow, and spent twenty-three minutes moving the distance of the twenty-three-foot tether with the help of an oxygen gas jet maneuvering gun. The spacewalk, called an *extravehicular activity* (EVA) in NASA-speak, was so exhilarating that when mission control ordered him to return to the capsule, White was reluctant. "It's the saddest moment of my life," he said as he reentered the spacecraft.

The handheld maneuvering unit (HHMU) White used to swoop around the Gemini capsule had a mounted camera affixed to it and was primitive, to say the least: It exhausted its pressurized gas after only three minutes. White's partner McDivitt actually deemed it a failure, because unless it was precisely aimed relative to his center of mass, all it did was cause him to rotate rather than move forward. Clearly, some engineers had forgotten about the challenging physics of the situation and assumed you could simply point the unit in the opposite direction of your motion and pull the trigger.

Similar units were used in later missions, but they were eventually replaced by more robust, strap-on manned maneuvering units (MMU), first used in the Space Shuttle program starting in 1984. Astronauts Bruce McCandless, a codeveloper of the MMU, and Robert L. Stewart used MMUs for their historic untethered EVAs on February 7, 1984. A photograph of McCandless during one of his EVAs, taken when he was more than three hundred feet from the shuttle *Challenger,* has become one of the most recognizable photos of an astronaut in space.

So, while White's HHMU may have been less than fully functional, you have to start somewhere. His use of the device represents the first historic step along the path to the extraordinary mobility astronauts enjoy in space today.

◀ The famous image of Bruce McCandless on an EVA

70

Apollo 1 Block I Hatch

A grim wake-up call to the perils of space travel

1967

Only six years after Alan Shepard's successful suborbital flight, one of the most horrific accidents that could happen to a rapidly advancing space program took place. The Apollo program, initiated in 1963, was destined for glory and would go on to reach the Moon in 1969. But it began in tragedy. A deadly fire broke out during the first ground test of a fully functioning Apollo capsule, *Apollo 1,* in 1967. An electrical spark, combined with the atmosphere of pure pressurized oxygen and the presence of combustible nylon, turned the interior of the capsule into a burst of roaring fire for several seconds, enough time to kill the three astronauts inside: Virgil "Gus" Grissom, Edward White, and Roger Chaffee. The official accident report indicated that nine seconds prior to the fire, a momentary voltage increase on AC bus 2, a kind of electrical connector, had been detected. But they hadn't yet found the source of the surge.

The autopsies showed that the astronauts weren't killed by the flames but rather by asphyxiation from carbon monoxide. The accident review board identified a number of electrical arcs close to the spacecraft's Environmental Control Unit (ECU). In particular, a single silver-plated copper wire had become stripped of its Teflon insulation thanks to frequent opening and closing of a small access door. The wire ran close to an ethylene glycol $(C_2H_6O_2)$ cooling line that was prone to leaks. Simulations showed that it was entirely possible that these leaks could become an ignition point in a pure-oxygen atmosphere. Ethylene glycol is a common and effective coolant that's used in automobile radiator systems. As it circulates, it absorbs heat more efficiently than water, and it has a higher boiling point. It is used in space-suit cooling systems as part of an undergarment, and it also provides much-needed cooling in the cramped

◀ The Block I hatch as seen on a later mission (*Apollo 4*)

quarters of space capsules like Apollo. Which isn't a problem as long as it remains out of contact with both oxygen and an ignition source.

But perhaps the most consequential design failure was the hatch. Even if everything else had failed, an easy way to eject from the cabin could have saved the astronauts, at least in theory. But the Block I hatch used on *Apollo 1,* known as a plug hatch, relies on pressure being higher inside the cabin than outside in order to keep it plugged shut. Plus, it only opened inward, and only once the interior pressure had been reduced. Its design was meant to prevent accidental openings during the mission, but the *Apollo 1* tragedy exposed its fatal flaws. The vent valve that needed to be engaged to release pressure before the hatch could be opened was blocked by the wall of flames, and, in any case, the hatch wasn't designed for the kind of rapid pressure release necessary in such an emergency.

The Apollo space capsule was completely redesigned to avoid any future fire-related problems, starting with the hatch, which was made to open outward, without any need for pressure rebalancing first. The cabin air was changed to a mixture of 60 percent oxygen and 40 percent nitrogen; the pure nylon, highly flammable space suits were replaced with nonflammable fabrics; aluminum tubing was replaced with less reactive stainless steel, and wires were covered with flame-resistant insulation.

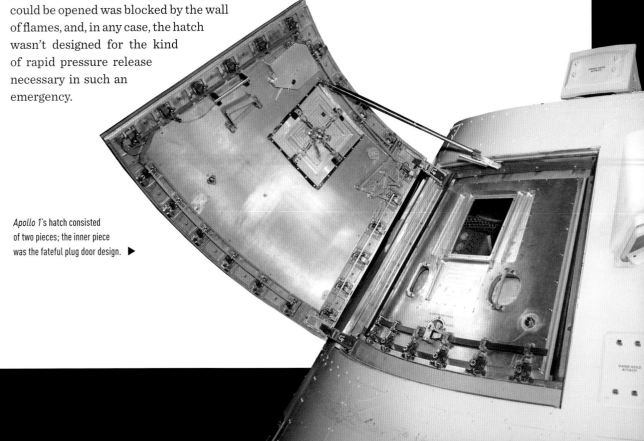

Apollo 1's hatch consisted of two pieces; the inner piece was the fateful plug door design. ▶

71

The Interface Message Processor

The beginning of the World Wide Web

1967

The Interface Message Processor (IMP) was a key technology developed for an experimental computer network known as ARPANET, the earliest iteration of what would eventually become the internet. An IMP served as a gateway—a small computer that connected a participant's computer with the ARPANET/internet backbone; today, these are called *routers*. Its role was to forward data packets from the host computer to other computers on the network using the TCP/IP data packet protocol, also known as "packet switching."

The first IMPs were developed in 1966 by Donald Davies on the NPL Network at England's National Physical Laboratory; and in 1967 by the ARPANET implementation team led by Larry Roberts, independently of Davies. Wesley Clark at Washington University in St. Louis came up with the idea that the IMP should be a small computer located between the host and the network. Four of these IMPs, based on the Honeywell DDP-516 minicomputer, were contracted to be built by the Massachusetts-based company Bolt, Beranek, and Newman in 1969. By the end of the year, the IMPs had been distributed to all the participating nodes. The first packet-switched message was sent on October 29, 1969, between Charley Kline at UCLA and Bill Duvall at Stanford Research Institute. It consisted solely of the letters *LO,* which was supposed to be *LOGIN*, but the network link crashed. Incidentally, a version of the DDP-516, the DDP-316, was marketed as the Honeywell Kitchen Computer, "useful for storing recipes," at a cost of $73,000 in today's dollars.

For many years, astronomers would use ARPANET connections through their local mainframe computers to transfer data and messages. By the end of 1972 there were twenty-four nodes on ARPANET, including NASA and the National Science Foundation. Email appeared in 1971, followed by the File Transfer Protocol (FTP) in 1973. You could even remotely log on to a computer using the new Telnet service by 1972. ARPANET was eventually decommissioned in 1990 and replaced by NSFnet. All of this technology innovation led to the birth of the World Wide Web, invented by English computer scientist Tim Berners-Lee in 1989. He wrote the first web browser, called *World Wide Web,* in 1990 while employed at CERN, the particle physics laboratory near Geneva, Switzerland.

Because it facilitates the rapid exchange of information and data via computer interfaces available in nearly every astronomer's office, the internet is a vital feature of all astronomical research today. Moreover, the same technology that has made the internet universally available has also supported developments that have dramatically increased the speed of computers used to operate spacecraft and to perform a variety of mathematical modeling calculations.

INTERFACE
MESSAGE
PROCESSOR

Developed for
the Advanced Research Projects Agency
by Bolt Beranek and Newman Inc.

bbn

148

72

The Hasselblad Camera

The first selfies in space

1968

◀ *Earthrise* photo taken by Bill Anders on *Apollo 8*

I n October 1962, astronaut Walter Schirra brought a Hasselblad 500C camera into space with him on the Mercury mission and snapped the first space photographs. Since then, Hasselblads have been the NASA space program's favorite still-frame cameras. Manufactured by the Victor Hasselblad AB company, based in Gothenburg, Sweden, the medium-format cameras were used extensively in the Gemini, Apollo, and Skylab programs, and some of the most iconic images of outer space have been taken with them.

The Model 500EL accompanied all of the Apollo missions, and a dozen of the boxy silver cameras have been left behind on the lunar surface—only their crisp, clear negatives were brought back to Earth. On Christmas Eve 1968, *Apollo 8* astronaut William Anders took the famous image entitled *Earthrise*

with a 70 mm Hasselblad 500EL fitted with a 250-mm telephoto lens. He had just completed an extensive photography session of the lunar surface and the camera was loaded with a fresh film cartridge: custom Ektachrome film developed by Kodak. As with all of the Hasselblad cameras used by NASA, Anders's 500EL was specially designed for use in space. Its silver color helped with thermal control when exposed to the space environment. It came with a special glass Reseau plate, which has a grid of crosses etched into it to aid in determining the geometry of the scenery and estimating distance. The lenses were also precisely calibrated to remove field distortion.

In a frantic few seconds, Anders recognized the beauty of the evolving scene he was seeing from the window of the Apollo command module and quickly snapped a sequence of stills, of which *Earthrise* was the most perfectly focused and artistically framed. More than just a lovely photo, it went on to capture the imagination of millions of humans back on Earth, providing inspiration for the budding environmental movement and future Earth Day celebrations.

◀ Hasselblad 500EL/M used in the Apollo program

73

Apollo 11 Moon Rocks

The first systematic geological samples from another world

1969

What, exactly, is the Moon made of? This question has been asked for millennia, but after the advent of the telescope in 1609, the answers began to be based more on reality than speculation. Many indirect methods of observation were developed, such as measuring the reflectivity of the Moon, but all that could be said was that it must be made of rock of some kind. The first hard evidence of its geological composition came by means of astronauts actually visiting the surface and seeing for themselves—the lunar rock and debris they brought back was more definitive than centuries of speculation could ever be.

The twenty-one hours and thirty-six minutes that Neil Armstrong and Buzz Aldrin spent on the lunar surface included a single two-and-a-half-hour EVA in which they collected around forty-eight pounds of rock and soil samples. They also deployed several experiment packages as they raced against the clock on a complicated time line of carefully scripted activities.

Among the rock samples was No. 10072,80, identified as "vesicular, fine-grained, high-potassium, ilmenite basalt." This 447-gram rock was dated to be 3.6 billion years old, but its cosmic ray dating age is only about 235 million years. This means it spent most of its life buried, but 235 million years ago it found its way to the surface, perhaps because of a meteor impact explosion in the lunar regolith. A new mineral was found among the *Apollo 11* rocks, one not previously found on Earth. It is called *armalcolite*, after the *Apollo 11* astronauts: ARM-strong, AL-drin, and COL-lins.

This and other rock samples demonstrated that the lunar surface is rich in silicate, just like Earth's crust. However, the abundance of titanium and aluminum oxides, which combine to account for 20 percent of lunar samples, means that the Moon was not formed solely from material similar to Earth's crust. Somehow, the formation of our Moon was different from that of Earth. This new information led to the giant-impactor theory, which proposes that a large Mars-sized body collided with Earth, and its material mixed with Earth's ejected crust to create the matter out of which the Moon formed.

Modern CCD imager

▼

◄ Historic image of Pluto's moon Charon taken with the New Horizons Long Range Reconnaissance Imager (LORRI) digital camera on July 14, 2015, when New Horizons was at a distance of three billion miles from Earth

74

The CCD Imager

Filmless imaging of planets, stars, and galaxies

1969

In 1961, Eugene Lally at the Jet Propulsion Laboratory showed how digital imaging technology, which did not yet exist, could theoretically be used to define a spacecraft's orientation in space and its current trajectory. In fact, Lally coined the term *digital photography*. The word *pixel* was in use by 1965 and is a combination of the words *picture* and *element*.

There were several attempts at creating a tubeless imaging system in the 1960s. The charge-coupled device (CCD), the first true solid-state device with all the modern features of a digital imaging system, was developed in 1969 by AT&T Bell Laboratories engineers George Smith and Willard Boyle, who went on to receive the Nobel Prize for this work. Ironically, they developed it as a memory device to replace magnetic bubble memory in the design of a "picture phone." It was only a year later that engineers discovered that the semiconductor material in these CCDs was light-sensitive, meaning CCD memory units could also work as imagers. Very quickly, a number of companies began developing CCD arrays and experimenting with them for digital camera applications, with Fairchild Semiconductor taking an early lead.

Even though the Fairchild Semiconductor's CCD201ADC, a 100-by-100-pixel array, could "take" a still image and store it in the accumulated charges in each picture-cell (pixel), the image would fade quickly. That's why the early applications of CCD technology in the 1970s were in the design of solid-state TV cameras such as the Fairchild MV-100. In 1973, the young engineer Steven Sasson went to work for Eastman Kodak and solved this image-fading problem by designing a circuit that would write these numbers onto magnetic audiotape. What was so earth-shattering about this humble start is that the entire process was electronic—no chemicals needed. Also, the camera could be "framed" for longer times to register faint scenery. It took twenty-three seconds for a photo taken with Sasson's 0.01-megapixel camera to form on the CCD. (In 2017 alone, more than 1.2 trillion digital photos were taken, mostly on smartphones—and mostly, it seems, of people's pets.)

The first application of digital photography in astronomy came a few years later, in 1976, when University of Arizona astronomer Bradford Smith attached a Texas Instruments 400-by-400-pixel CCD to the sixty-inch telescope on Mount Lemmon and took the first grainy images of Uranus. These revealed details of the planet's atmosphere for the first time.

NASA quickly adopted CCD imagers for spacecraft, using them not only to photograph planetary surfaces, but as a key component in their navigation systems. CCD imaging has completely transformed both space exploration and astronomy and continues to be instrumental: It will be used in the largest digital camera ever built, a three-gigapixel imager within the Large Synoptic Survey Telescope, set to go operational in 2022, which will be capable of taking high-resolution images of the entire night sky.

75

The Lunar Laser Ranging RetroReflector

Measuring the Earth–Moon distance with lasers

1969

◀ The *Apollo 15* corner cube retroreflector on the lunar surface

How far away is the Moon? This basic question has been debated for centuries; by the 1950s, we could use radar to establish a distance precise to within a few miles. But the advent of the laser, and the phenomenally precise metrology, or technique of measurement, that it promised, offered a whole new universe of ways to determine this important and basic astronomical number. But to accurately calculate the distance would require nothing less than a manned mission: the NASA Apollo program.

Simply placing a mirror on the lunar surface wouldn't suffice to facilitate laser measurement, because the astronauts' cumbersome space suits and manual orientation techniques could not guarantee the accurate alignment needed to reflect a pulse of laser light back to the telescope from which it originated. But the advent of radar during World War II had given us a new reflection technique in which the optical signal would be returned to the sender regardless of the orientation of the reflector. Called the *corner reflector*, this approach consists of three mirrors located at the corner of a cube, with their reflective surfaces facing inward. Any signal approaching this reflector from any angle inside the field of the corner cube will be reflected exactly back along the direction of entry.

In 1964, only a few years after the invention of the laser, NASA received the first laser return from the *Explorer 22* satellite, which was equipped with a corner cube reflector. This technology was quickly adapted to become one of the main experiment packages that *Apollo 11, 14,* and *15* astronauts transported to the lunar surface beginning in 1969: the Lunar Laser Ranging RetroReflector. Although a powerful laser beam starts out only a few millimeters wide on Earth, by the time it reaches the lunar surface it is roughly four miles in diameter. Of the one hundred thousand trillion photons emitted by the laser from Earth, only one photon will be returned every few seconds; these are detected by a large telescope using a sensitive photometer. The Apollo experiments finally gave us the answer we'd been searching for: The Moon's distance from Earth varies over the course of a year, but its average distance is around 238,855 miles. And with their down-to-a-millimeter precision, these instruments also made another remarkable discovery: Our Moon is moving away from Earth at a rate of about an inch and a half per year.

76

The Apollo Lunar Television Camera

The iconic image of Neil Armstrong's first, small step

1969

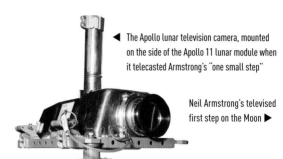

◀ The Apollo lunar television camera, mounted on the side of the Apollo 11 lunar module when it telecasted Armstrong's "one small step"

Neil Armstrong's televised first step on the Moon ▶

On July 21, 1969, at 02:56:15 UTC, astronaut Neil Armstrong placed his foot on the lunar surface, becoming the first human to set foot on another celestial body. As he climbed down the nine-rung metal ladder, he pulled a ring on a lanyard to deploy an equipment stowage box and activate the TV camera. You would think that the video documentation of this historic footstep would have been of the highest quality, but in fact it was crude and hard to visually understand when it was finally broadcast to more than six hundred million people back on Earth. The black-and-white video camera, built by Westinghouse, measured eleven by seven by three inches and weighed seven pounds. It used about seven watts to operate and was designed to function in the extreme lunar temperatures—from 250 degrees Fahrenheit in daylight to –251 degrees Fahrenheit in the shade. It also needed a slow-scan capability of about one frame per second because the lunar surface was very dark, as were the shadows. More important, the video signal had to be kept to a bandwidth of only seven hundred kHz so that it could be transmitted via the lunar module's S-band antenna.

Because of its unique design meant to accommodate the rigors of space, the lighting conditions on the Moon, and the S-band telemetry constraints, the slow-scan video camera was actually incompatible with commercial-grade TV. The video signal relayed to NASA's international tracking stations had to be displayed on a special monitor and then reimaged with a standard TV camera pointed at the monitor. The result was a rather low-grade video with ghostly persistent images of the astronauts. About twelve minutes after the historic video sequence was recorded, the astronauts placed the video camera on a tripod and used it to take panorama sequences of the landing area. Meanwhile, the historic photos from the *Apollo 11* landing site were all taken with a professional-grade Hasselblad still camera, which the astronauts used to take hundreds of high-resolution pictures of this alien landscape.

Crude as the video was, the film still that captures the moment of Armstrong's first contact with the Moon remains one of the most iconic images in our entire history of space exploration. Technology is always becoming sharper and more refined, but there's something to be said for the earliest depictions of our first steps into new realms of understanding and discovery—such as the ancient Nebra sky disk or the early daguerreotype photos of the Sun and Moon—however imperfect those depictions may be.

77

◀ Deep underground, a researcher stands over the giant perchloroethylene tank that makes up the neutrino detector.

The Homestake Gold Mine Neutrino Detector

The first neutrino detector

1970

Electrons, neutrons, and protons are the familiar stuff of modern nuclear physics, but even before the neutron was discovered in 1932, studies of radioactive decay in the 1920s suggested the existence of another kind of subatomic particle. It was known that certain radioactive isotopes decayed into stable nuclei but emitted an electron in the process. For example, radioactive carbon-14 decays into nitrogen-14 and emits an electron. The nuclear transaction requires that one of the neutrons in the carbon nucleus be transformed into a proton in the nitrogen nucleus. This beta-decay process was investigated by Wolfgang Pauli in 1930 and led to the idea that new particles, later called *neutrinos*, had to carry away the missing energy from the decay process.

Once the origin of solar energy was identified as the thermonuclear fusion of hydrogen, scientists quickly recognized that the Sun should be a powerful source of neutrinos. Detecting them would be yet another test of whether the hydrogen fusion process was the source of solar and stellar energy. In the late 1960s, calculations performed by astronomer John Bahcall and an experiment designed by Raymond Davis led to the creation of a unique neutrino detector.

The detector consisted of a one-hundred-thousand-gallon tank filled with a common dry-cleaning fluid, perchloroethylene. The tank was buried nearly one mile underground in the Homestake gold mine, located in Lead, South Dakota. When a neutrino interacted with the fluid, one of the chlorine-37 nuclei would transform into an argon-37 nucleus after absorbing the solar neutrino. The radioactive argon atoms would be collected and counted, giving a measure of the number of solar neutrinos entering the tank every second. Several years of counting led to only one-third of the expected number of neutrinos, which forced the physics community to look deeply into what they thought they knew about neutrinos. In 2001, the process of neutrino oscillation was discovered, which explained how neutrinos from the Sun were changing en route to Earth into other kinds of neutrinos. This was why Davis's experiment could not detect all the expected neutrinos. When a full accounting was performed that included these two other types, the muon and tauon neutrinos, the discrepancy was eliminated.

Why all the fuss about neutrinos? Because they are a way of seeing and understanding how stars produce their energy and of investigating the earliest instants after the formation of our universe fourteen billion years ago. Davis's work on neutrinos would eventually earn him the Nobel Prize in Physics in 2002.

Today there are a number of different neutrino detectors in operation, and on the more distant horizon are detectors that will be able to explore the cosmic neutrino background left over from the Big Bang itself.

78

Lunokhod 1

The first robot to visit another world

1970

Robotically exploring a planet, moon, or asteroid is always less expensive than sending humans into space—we require food, water, air, and a pressurized spacecraft. If robotic technology can be perfected, there is no limit to the kinds of objects we can explore at a fraction of the cost of a manned program.

Shortly after NASA's successful Apollo program put astronauts on the Moon, the Soviet Union made its own lunar mark by launching the *Luna 17* mission to the Moon on November 10, 1970. Inside was a rover called *Lunokhod 1*, which was to disembark from the lander and spend the better part of a year dispatching scientific findings along its nearly seven-mile journey. It was followed in 1973 by *Lunokhod 2*, which survived for four months and traveled twenty-three miles, returning over eighty thousand images. In 1993, *Lunokhod 2* was sold at Sotheby's auction house in New York for $68,500, becoming the first privately owned spacecraft in the solar system.

These rovers were bathtub-shaped, about seven feet long, and contained radioisotope heaters to keep them warm at night. Their eight wheels were independently maneuverable, but special lubricants had to be used to keep the gears and motors functioning under the vacuum conditions and wide swings in temperature. In many ways, roving on the Moon is far harsher than similar operations on Mars. The final resting places of the two rovers remained a mystery until NASA's Lunar Reconnaissance Orbiter mission, whose job was to image the lunar surface at a resolution of six feet (two meters), located them via a series of images taken in 2010.

Since the Lunokhods stormed the lunar surface, four rovers have successfully landed on the surface of Mars, but the Moon wasn't visited again until the Chinese Yutu rover arrived on December 14, 2013. A subsequent *Yutu 2* rover landed on the far side of the Moon on January 3, 2019, becoming the first vehicle to operate there.

79

The Skylab Exercise Bike

Humans learn how to stay fit in space

1973

◀ Skylab

Contrary to decades of science fiction lore, living and operating in space is not without its hazards, which go far beyond meteor storms and solar flares. Following three million years of evolution on Earth, humans are not adapted to function in space under microgravity conditions. This news really began to hit home between 1965 and 1967, when astronauts on the *Gemini 4, 5,* and *7* missions, and later the *Soyuz 9* cosmonauts in 1970, were found to be suffering from a form of mild bone loss now called *spaceflight osteopenia.* In addition to bone loss, other physiological effects were detected, such as blood settling into the upper body and prolonged instances of vertigo and nausea induced by "space sickness."

Launched into orbit on May 14, 1973, Skylab was the first attempt at setting up a robust manned laboratory in low Earth orbit. The intent was to send rotating teams of astronauts up to Skylab via Saturn IB boosters and to do detailed medical research on their process of adaptation to the microgravity environment. Prior to Skylab, the longest successful manned flights were *Gemini 7* (fourteen days) and *Soyuz 9* (eighteen days). The third Skylab crew remained in the station for eighty-four days until the mission ended in February 1974. So Skylab was outfitted with an exercise bike and a Super Mini-Gym, a kind of centrifugal exercise machine, which the astronauts were encouraged to use to maintain fitness.

The scientific returns from Skylab were enormous, not only in terms of studies of the Sun but also in terms of knowledge gained about the effects of prolonged stays in space on humans. Pioneering advances were made in the understanding of the effects of bone loss, the pooling of blood in the upper extremities, and the impact of vigorous exercise on these conditions.

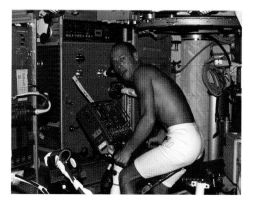

◀ Astronaut Pete Conrad on exercise bike in Skylab

80

The Laser Geodynamics Satellite (LAGEOS)

Discovering the true shape of Earth

1976

Back in the seventeenth century, Isaac Newton first proposed that Earth is not a perfect sphere, and in the eighteenth century, he was proved right. Instead, our planet is contorted in complex ways by the gravitational pull of the Sun and Moon into an oblate spheroid—critical information for navigators, who needed more accurate maps of the globe in order to travel the seas and reach safe harbor. During the nineteenth and twentieth centuries, there was an entire scientific industry devoted to measuring the shape of Earth with greater and greater accuracy. This turned out to be almost impossible to do by surveying the land and sea directly from Earth's surface. With the advent of space exploration and satellite technology in the 1960s, however, the geometrology game was revolutionized.

As a satellite orbits Earth, it experiences slightly different gravitational tugs along its orbit, which cause the satellite to lower or raise its orbit altitude slightly. By keeping track of these altitude changes using the timing of radio or laser signals, the shape of Earth's surface can be deduced.

The first satellite dedicated to performing these measurements was the Laser Geodynamics Satellite (LAGEOS), launched in 1976. It was followed in 1992 by LAGEOS-2, launched by NASA in collaboration with the Italian Space Agency. Both spacecrafts were aluminum-covered brass spheres with a diameter of twenty-four inches and masses of 897 and 893 pounds, respectively. Each was covered with 426 cube-corner retroreflectors, giving them the appearance of giant golf balls. From the ground, a laser pulse would be aimed at each satellite and reflected back to the ground station, where the timing difference was converted into an altitude, since light travels at a known speed of 186,000 miles per second. From millions of these one-shot measurements, Earth's shape could be determined to within a few centimeters. This provided enough information to prove with tremendous accuracy that our planet is indeed flattened at the poles and bulging at the equator.

But the data also revealed something even more interesting: There are further irregularities in Earth's gravitational field throughout the continental land masses and oceanic basins, which, when modeled, result in a rather unrecognizable shape that has come to be known as the Potsdam Gravity Potato, named after the research team in Potsdam, Germany, that created the visual representation. The minute altitude differences are magnified by thousands of times and used to create an exaggerated globe of Earth that emphasizes these irregularities.

◀ The Potsdam Gravity Potato: A visualization of Earth's gravity field produced by the GFZ German Research Center for Geosciences in Potsdam, Germany

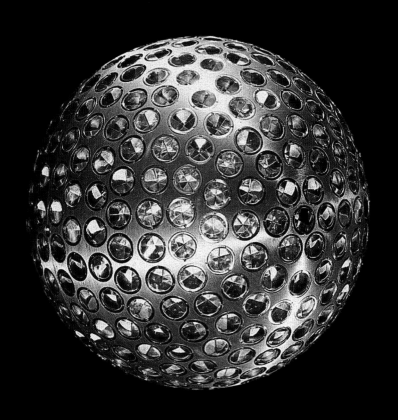

81

Smoot's Differential Microwave Radiometer

Big Bang cosmology confirmed

1976

Between 1976 and 1978, astrophysicist George Smoot tried to detect the Doppler motion of Earth as it moved through space with respect to the cosmic microwave background radiation. This radiation is the light left over from the Big Bang itself, now only visible at radio wavelengths. His instrument, called a *differential microwave radiometer* (a device that measures electromagnetic radiation in microwave wavelengths) was mounted on a high-flying U-2 spy plane in 1976. This historic study—revealing the universe's uniform expansion and its lack of rotation, among other things—and the design of the radiometer were the precursors of the NASA Cosmic Background Explorer (COBE) spacecraft, launched in 1989 to map and analyze the light left over from the Big Bang itself.

The instrument Smoot used was a very sensitive microwave radio receiver with two windows

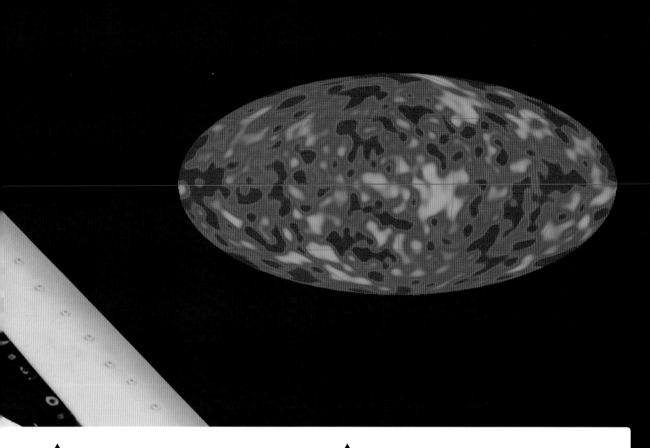

▲
The historic differential microwave
radiometer mounted on a U-2 spy plane

▲
Map showing irregularities in the
cosmic microwave background

that looked at the sky from two locations exactly sixty degrees apart. Two viewing frequencies, of 33 GHz and 54 GHz, were observed via two pairs of horn antennas. The larger pair, at 33 GHz, was selected to minimize the emission from the galaxy filled with interstellar gas and ionized plasma. The smaller, higher-frequency pair operating at 54 GHz monitored irregularities in Earth's upper atmosphere as the plane cruised higher than sixty-five thousand feet. The difference in intensity between the two viewing directions allowed many sources of local interference to be canceled out, leaving behind only the weak leftover signal caused by the Milky Way's movement through space.

This instrument was successful in measuring the Milky Way's motion relative to the cosmic microwave background. Building on this, the COBE spacecraft used an improved, more sophisticated version of the same technology: It was also called the differential microwave radiometer (DMR), and it succeeded in mapping irregularities in the cosmic microwave background. The all-sky DMR map was itself a groundbreaking verification of the Big Bang theory. A similar instrument was flown on NASA's Wilkinson Microwave Anisotropy Probe (WMAP) mission and the European Space Agency's Planck satellite, ushering in a new era of high-precision cosmology.

82

The Viking Remote-Controlled Sampling Arm

Robotic manipulation of another planet's surface

1976

Viking Mars lander surface sampler collector head, housed in the archives of the NASA Langley Research Center in Hampton, Virginia ▼

Since the Surveyor program's first Moon landing in 1966, scientists have wanted to poke and prod planetary surfaces to dig trenches for mineral samples or capture specimens for on-the-spot chemical analysis. It is not a trivial task to design a robotically controlled arm with motors and lubricants that can withstand the rigors of space. For the Surveyor landers, all that was needed was a simple scoop on a retractable arm. Even if this mechanism failed, the primary mission would still be largely successful simply by landing a piece of hardware on the lunar surface without crashing it and photographing the surroundings. But for the Viking program of 1976, the goals were far more complex and demanding. The entire success of the mission rested on the extendable arm and scoop being able to deliver a few ounces of Martian soil to an onboard chemistry laboratory to search for exotic molecules related to organic life, which every science fiction author expected to find there.

The first part of the arm, called a Surface Sampler Acquisition Assembly (SSAA) in NASA-speak, was simply a mechanism that extended the arm up to ten feet to reach a spot imaged in advance by the camera system. Once

there, a multipurpose scoop called the Surface Sampler Collector Head, about eight inches long, would be used to dig a trench and capture several ounces of pristine subsurface soil. This sample would then be quickly maneuvered over to the entrance port of the chemistry lab. The sampler would be rotated and shaken to sieve smaller grains into the lab for processing. Many things had to go exactly right in order to make this historic measurement; dozens of subsystems were required to work automatically and in the correct sequence.

In the end, the Martian soil showed no evidence of organic material. Subsequent studies of the data from the chemistry lab have, however, called this assessment into question. The *Curiosity* soil sampler, using the legacy technology developed by the Viking sampler, took the SSAA one huge step further by replacing its sampling head with a collection of instruments for drilling and chemically studying the sample on the spot. The results of hundreds of samples seem to show that the surface chemistry is more complex than previously thought, and that evidence of organic processes cannot be entirely eliminated—at least for now.

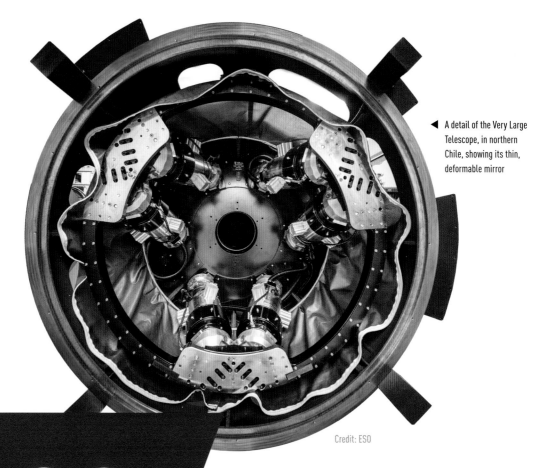

◀ A detail of the Very Large Telescope, in northern Chile, showing its thin, deformable mirror

Credit: ESO

83

The "Rubber Mirror"

Advent of adaptive optics, a better way for telescopes to see

1977

It has been the bane of astronomical research for centuries: When starlight passes through Earth's shifting and turbulent atmosphere, the stars appear to twinkle. This twinkling represents the point-like image of a star shifting its location in the sky by arc seconds from millisecond to millisecond, in a random dance we perceive as twinkling. Because of this distortion, when even the most perfectly designed telescope looks at stars, or the minute features on the Moon, or the surface of Mars, the images shift and blur, and photographs turn out fuzzy. There are three ways to overcome this effect: Two are difficult, and the third one is relatively easy.

The easy method, at least in principle, is to take photographs in quick succession so that each image follows the celestial object as it shifts position. This can require taking hundreds of images every second. Then you throw out the bad images, save the good ones, and combine the good images by recentering each one to the same location, thereby canceling the twinkling movement. Because the photographic

These images of the planet Neptune were obtained with the MUSE/GALACSI instrument on the European Southern Observatory's Very Large Telescope. ▶

Credit: ESO/P. Weilbacher (AIP)

exposures are very short, this method works only for very bright objects like the Sun, Moon, and a few of the brighter planets: Fainter objects such as nebulae and galaxies do not yield enough light. Although this easy method, called *scintillation suppression*, has never worked that well in practice, the two difficult methods have performed with spectacular results.

The first of these is demonstrated by the Hubble Space Telescope: Place the telescope high above Earth's troublesome atmosphere. This is expensive, though, and is limited to telescopes that are quite small compared to the giants now in operation on the ground.

The second difficult method uses adaptive optics and was first proposed by astronomer Horace Babcock in 1953. When light passes through the turbulent atmosphere, those portions of an image that enter a telescope can be shifted in phase because they travel on slightly different paths to get to the point of focus. This phase shifting across the image is what causes the optical blurring. Babcock suggested using a mirror that could be deformed in order to counterbalance any distortion.

The US military had a hand in the early development of Babcock's theory, but a key breakthrough came from Nobel-winning experimental physicist Luis Alvarez and his team at the Lawrence Berkeley National Laboratory. In 1977, they constructed a so-called rubber mirror, like the one Babcock had proposed, to correct an image. The team, including physicist Frank Crawford, developed a working model of a telescope with this flexible "rubber"

mirror, proving in principle that the concept could be used when viewing stars.

It was an idea ahead of its time: Twenty years would pass before technology would catch up with the concept with enough precision to be of value to astronomy. Today, adaptive optics builds on Babcock's concept by using lasers to create a "star" in the sky whose phase properties are exactly known. When a perfectly focused image is formed, the arrival of the electromagnetic wave is in-phase across the entire image. Atmospheric scintillation causes the phase of different portions of the arriving wave to be out-of-phase because of the different paths taken through the atmosphere. The laser star is created by a perfectly in-phase source so any changes to its phase across its image are due to atmospheric scintillation. This phase information from the laser star is then used a thousand times a second to physically adjust the shape of the smaller secondary mirror in a telescope, canceling out the phase errors across the desired astronomical image. What results is a perfectly in-phase image corrected for atmospheric twinkling. The technique can be used on both bright and faint objects, and allows a telescope to perform to its full optical ability.

The use of adaptive optics has become standard: All major astronomical observatories now employ this technique, and because their mirrors can be far larger than the one in the Hubble Space Telescope, ground-based telescopes routinely out-perform Hubble in certain kinds of investigations, and at a far lower cost.

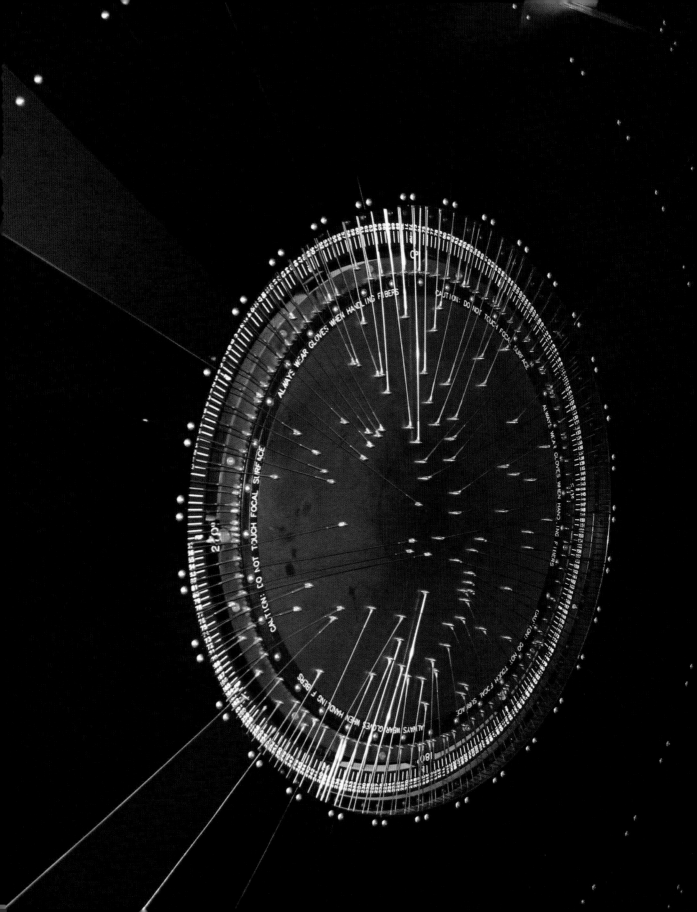

Detail of spectra of about one hundred stars, imaged using multiple fibers to pipe light to a single slit in the Hydra optical wavelength spectrograph

Credit:ESO

◀ The National Optical Astronomy Observatory's Hydra multi-fiber spectrograph

84

The Multi-Fiber Spectrograph

Studying galaxies a hundred at a time

1978

For decades, astronomers had to gather high-resolution spectroscopic data (maps of dispersed electromagnetic radiation) on objects one at a time. Some systems, when at a lower resolution, let you photograph an entire field of stars in one long exposure, but to resolve the fine details of a spectrum for chemical analysis, you had to pass the starlight individually through the spectroscope. The challenge was to get the light from a number of single objects to enter the spectrograph in such a way that their light did not intermingle. This problem was solved in the 1970s thanks to the development of high-quality optical fibers for the communication industry.

Astronomer Roger Angel and his graduate students at the University of Arizona tested out the quality of these new optical fibers in 1978 by piping the light from the quasar 3C273, viewed with a thirty-six-inch telescope, through a twenty-meter fiber and into the spectrograph. The results were extremely promising. A year later, the first twenty-fiber spectrograph, called Medusa, was created by positioning each fiber to align with the locations of eight galaxies within the cluster Abell 754.

During the early years of this technology, an astronomer had to photograph the area of sky that caught their interest and, on a metal plate, drill a hole at the location of each object to be studied at exactly the same scale as the photograph taken at the focal plane of the telescope. This time-consuming process was usually done several weeks before the observing session. Then the plate had to be mounted into the focal plane of the telescope, and a single fiber-optic cable was glued or mechanically affixed into each hole. The cable was then connected to the telescope's spectrograph. This tedious method was called the *plug-plate technique*, but the end result was that you could simultaneously study the spectra of dozens of objects. By the 1980s, much of this tedium was automated, and today large observatories have these multi-object spectrographs ready for use as facility instruments, capable of observing four hundred or more objects at the same time each night.

Without this historic advance in spectroscopy, none of the galaxy surveys of the last two decades could have been attempted. One of the most daunting surveys, called the Sloan Digital Sky Survey, was initially undertaken between 1998 and 2008 using the Apache Point Observatory in New Mexico. The 2.5-meter telescope was equipped with a spectrograph capable of observing 640 galaxies simultaneously, and nearly 6,000 galaxies every night. The project continues today and has detected hundreds of millions of celestial objects, providing major breakthroughs in our understanding of the structure of the deep cosmos.

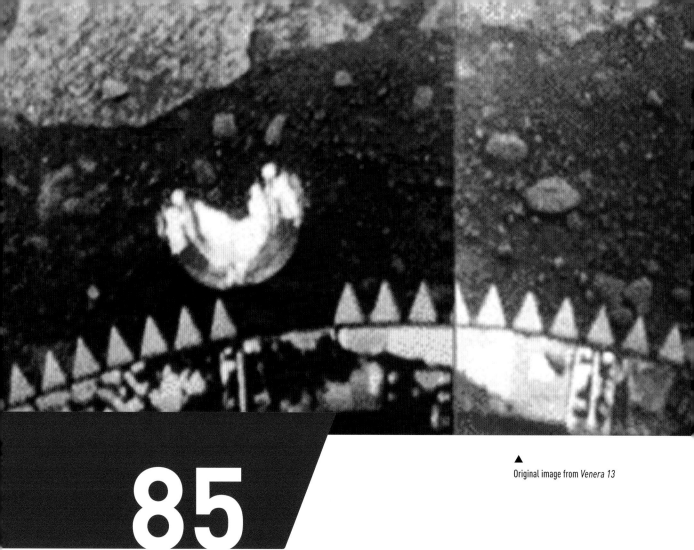

▲
Original image from *Venera 13*

85

The Venera Landers

Exploring the surface of Venus

1981

Model of a
Venera lander ▶

etween 1961 and 1983, the Soviet Union launched a total of sixteen spacecraft to study Venus—a series of orbiters, atmospheric probes, and landers. Several of these missions—from *Venera 7* through *14*—succeeded in landing an instrument payload on the planet's surface. *Venera 13* survived the longest, for more than two hours, before failing due to battery exhaustion. In 1966, *Venera 3* crashed into Venus, becoming the first human-made probe to impact another planet. *Venera 5* and *6* each relayed atmospheric data for a little less than an hour during their parachute descents. In 1970, *Venera 7* reached Venus's surface

and transmitted twenty-three minutes of data—it was the first probe to land on another planet. *Venera 9, 10, 11,* and *12,* launched between 1975 and 1978, were massive landers, weighing between two and five tons apiece. *Venera 9* and *10* both took photographs and survived for about an hour. Both *11* and *12* transmitted data for about a hundred minutes, but their cameras' lens caps malfunctioned, so no images were sent. The final pair of landers, *Venera 13* and *14,* were launched in 1981. They survived for 127 and 57 minutes, respectively, and took pictures of the surface. Because the surface temperature of Venus is over eight hundred degrees Fahrenheit and the atmospheric pressure is ninety Earth atmospheres (1300 psi), none of these Venera landers were expected to operate for more than thirty minutes or so before overheating. Nevertheless, each one contributed to our understanding of the planet, particularly the scope of the engineering challenge it presents.

The distorted pictures returned by *Venera 9, Venera 10, Venera 13,* and *Venera 14* revealed tantalizing glimpses of rounded pebbles and flat rocks out to the horizon, although from the perspective of the cameras, which mostly looked downward, that meant the apparent horizon was no more than a few dozen meters from the landers.

The technological challenges of operating on Venus are tremendous. The Venera landers were based on the conventional electronics of the 1960s and '70s, which involved silicon integrated circuits. These devices begin to malfunction at temperatures over 250 degrees Celsius. Once on the surface of Venus, the landers could not be actively cooled because they would require refrigeration systems that are heavy and that require additional sources of energy. Batteries also fail when overheated. NASA has recently developed silicon-carbide-based electronics and wiring that can function in Venus's extreme environment for many days. In the future, when we decide to return to Venus, the new-generation landers will perform dramatically better than the outmoded Veneras. And there's no telling what they may find with a little extra time on their hands.

86

◄ The black rubber O-ring within the space shuttle's solid rocket booster

The Compromised *Challenger* O-Rings

A humble sealant causes a historic disaster

1986

It's unclear who invented the first true round rubber ring known today as the O-ring. The earliest patent for it was registered in Sweden by J. O. Lundberg in 1896, and yet Thomas Edison may have been using his own version, which he called an *elastic stopple* in his electric light designs as early as 1882 (Patent No. 264,653). Credit also often goes to the Danish machinist Niels Christensen, who filed his own American patent (No. 2,180,795) for this device in 1937. He had been looking for ways to make hydraulic seals to use with metal pistons, and by trial and error discovered that a rubber ring-shaped gasket, when greased and compressed under pressure, did the trick perfectly.

Whoever first designed them, O-rings are everywhere today; they're used in garden hoses and faucets as well as spacecraft design and nuclear physics. The smallest O-ring manufactured is only 0.004 inches (0.1 mm) in diameter, and is used in medical devices and instruments. Among the largest O-rings are those used in the rocket industry to seal segments of solid rocket boosters, which can have diameters of up to twelve feet or more. Normally, they operate flawlessly, and no one pays much attention. Only when they fail do they become noteworthy. By far the most consequential failure of an O-ring came on January 28, 1986, the day of the worst disaster in spaceflight history: the space shuttle *Challenger* accident.

On the day the *Challenger* launched, the air temperature was well below the operating limits of the giant O-rings that sealed the connections between the segments of the solid rocket boosters. One of the O-rings lost its flexibility in the cold, allowing the seal that contained the inferno of combustion taking place inside the booster to rupture. The flames escaped and burned through the fuel tank, creating a powerful fireball. In the process, the shuttle was separated from both its booster rockets and fuel tank. Unable to survive the surrounding aerodynamic forces, it broke into several large pieces. The crew compartment fell into the Atlantic Ocean. All seven astronauts on board were killed.

The tragedy reminds how extraordinarily complex spaceflight truly is: Every element, even something as seemingly mundane as a rubber seal, plays a vital role and must work properly for a successful mission. O-rings might not normally command a whole lot of attention, but space exploration depends as much on these simple pieces of rubber as on any other space-related object.

◄ The *Challenger* accident of 1986

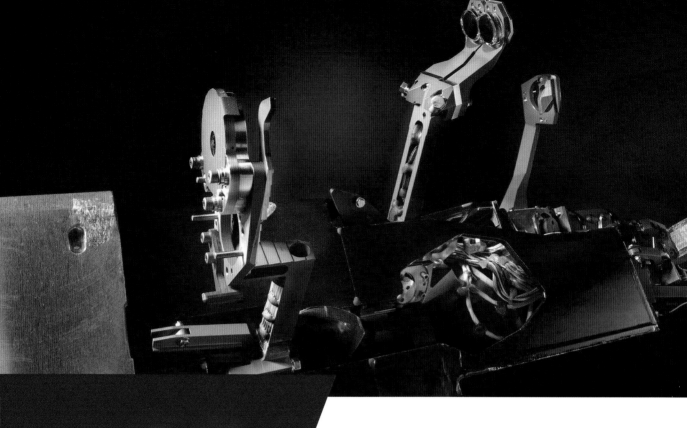

87

COSTAR

Giving the Hubble Space Telescope the gift of sight

1993

Astronomers have wanted a gigantic "mother of all telescopes" in space for decades. And not just any telescope, but an observatory-grade system with an aperture of at least a full three feet in order to study faint, distant objects in space. Earth's atmospheric distortion, the thing that makes stars twinkle, also robs all ground-based telescopes of their highest resolution, so space has always been considered the perfect arena for serious astronomical research. On April 24, 1990, astronomers got their wish. NASA launched the Hubble Space Telescope, with a massive ninety-four-inch mirror, through the aid of the space shuttle. It could be serviced every three years or so to fix any problems and swap in new instruments as the needs of astronomers changed and technology steadily improved.

Unfortunately, what was supposed to be the new gold standard of telescopes didn't work that well: The first images it returned were blurry. It quickly became apparent that Hubble could not be properly focused no matter how the mechanism was moved. But the

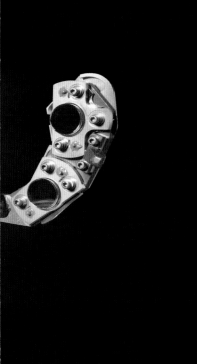

▲
COSTAR contained five pairs of small corrective mirrors on
deployable arms to send corrected light to other instruments
on the Hubble: the Faint Object Camera, the Faint Object
Spectrograph, and the Goddard High Resolution Spectrograph.

▲
Pillars of Creation, known to astronomers
as the Eagle Nebula, or Messier 16, is in the
constellation Serpens.

manner in which the blurring changed showed that the mirror was suffering from an optical defect called *spherical aberration* in which each annulus of the mirror was sending its focused light to a specific, but different, spot along the optical axis of the telescope. Some detective work revealed that the company responsible for fashioning the primary mirror had assembled the testing device with one of its lenses out of position by 1.3 millimeters. It was impossible to replace the primary mirror in orbit, so a new optical device called the Corrective Optics Space Telescope Axial Replacement (COSTAR) was built and flown up to the Hubble during the first space shuttle servicing mission (STS-61) in 1993. The telescope's original Wide Field and Planetary Camera was replaced by an updated version (WFPC2), which had COSTAR built in, and is responsible for some of the most beautiful images taken by the Hubble since then.

Thanks to these corrected optics, the number of major discoveries that have been made with the Hubble Space Telescope has been truly mind-boggling to contemplate, even for an astronomer. After operating for an astonishing twenty-nine years, it has captured over one million photographs of nearly forty thousand astronomical objects. It uses only 2,400 watts of electricity to run all of its systems—not much more than a small house. Every week it delivers around 150 gigabytes of data, mostly images, back to Earth. The famous image titled *Pillars of Creation* was taken by astronomers Jeff Hester and Paul Scowen on April Fool's Day, 1995, with the WFPC2. Without the corrective optics of COSTAR, Hubble wouldn't have been the famous telescope that it is today, but rather a failed telescope of seriously limited capability.

88

CMOS Sensors

High-precision astronomical imaging

1995

For decades, the favored solid-state imaging device was the charge-coupled device (CCD), invented in 1969. By the 1980s, CCDs were used extensively in commercial camcorders. The first digital camera, invented in 1975 by engineer Steven Sasson at Kodak, was, in fact, a CCD imager. Meanwhile, complementary metal-oxide semiconductor (CMOS) technology had been around since 1963 as a way to construct integrated circuitry for microprocessors, RAM memory, and other digital circuitry. As the consumer computer market grew dramatically in the 1980s, CMOS became the preferred technology when designing large RAMs with low power requirements.

These commercial developments set the stage for a team of engineers at NASA's Jet Propulsion Laboratory (JPL), led by Eric Fossum, to explore the potential of CMOS architecture and manufacturing techniques in the development of compact, low-power imaging sensors for spacecraft. The active pixel sensor, which he invented in the early 1990s, surpassed conventional CCD imaging systems in two ways: It offered both low power consumption and low noise, which allowed for cleaner images. And it had another major advantage: The image sensor could be manufactured on the same chip as the accompanying CMOS components, allowing one to create a miniaturized camera, with all necessary imaging technology, on a single chip. This dramatically saved manufacturing time and cost.

Despite Fossum's success in fabricating the first CMOS imaging sensor, NASA was still committed to CCD technology and did not pursue CMOS imagers further. Fossum quickly saw the enormous commercial potential for CMOS-based imagers, and he and his colleague at JPL, Sabrina Kemeny, decided to invest in the technology themselves. In 1995, they cofounded Photobit Corporation and licensed the CMOS technology from JPL. By 1998 they had designed and marketed their first digital camera chip, the PB-159. The PB-100 followed soon after; this second chip became the heart of the Intel Easy PC Camera, a webcam, and the Logitech QuickCam. These cameras brought PC videoconferencing into the mainstream and convinced the industry that CMOS imagers were the future. Photobit was eventually bought by Micron, and by 2013 more than one billion CMOS imagers were being manufactured every year. These days, they're a part of all smartphones.

Virtually all of NASA's many spacecraft missions continue to use CCD imagers, because CMOS sensors do not have the high level of imaging performance NASA needs. But CMOS sensors are a classic example of the way space technology can find its way into our everyday lives—in this case, the vast majority of us carry in our pockets imaging technology born from the space program.

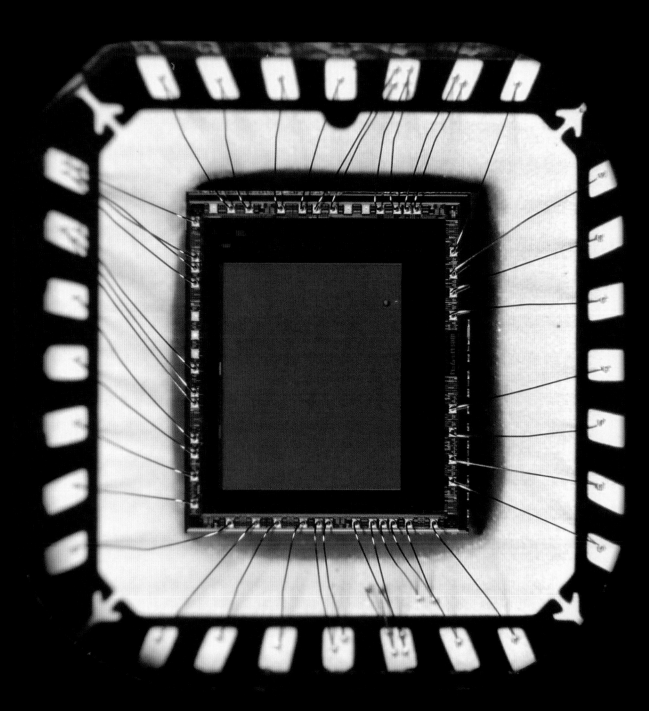

89

The Allan Hills Meteorite

The search for aliens gets serious

1996

This 4.5-billion-year-old rock, labeled
meteorite ALH84001, is one of ten rocks
from Mars in which researchers have found
organic carbon compounds that originated on
Mars without the involvement of biological
processes. ▶

On August 6, 1996, NASA scientist David McKay announced that he had found evidence for microscopic life on Mars in a meteorite called ALH84001, recovered from the Allan Hills area in Antarctica in 1984. What he found appeared to be small linear objects that resembled segmented nanobacteria. The announcement was so staggering that even President Clinton mentioned it in a press conference on the South Lawn of the White House the next day. But the discovery was met with considerable skepticism. To quote the late astronomer Carl Sagan: "Extraordinary claims require extraordinary evidence."

Intense investigation of the radioactive dating ages of various components in the

nearly five-pound meteorite pointed to a complicated origin on Mars. It had been launched from the surface of Mars about 17 million years ago, and eventually impacted with Allan Hills about thirteen thousand years ago, where it remained buried in the ice until it was exposed and then later recovered in 1984. Based on its mineralogy, it was more than 4 billion years old and had formed when there was considerable water on Mars. Around 3.6 billion years ago, some kind of fluid carrying carbonate minerals, probably water, found its way into the cracks of the igneous rock, forming or depositing a host of microfossils in the process.

The microfossils were only twenty to one hundred nanometers in diameter—far smaller than conventional virus particles containing DNA, but not smaller than viruses made from pure RNA. A number of subsequent investigations seemed to conclusively demonstrate that they were not biological in origin but could have been formed under natural geologic processes.

Although the evidence for life on Mars was not made conclusive by the discovery of this meteorite, it spurred a sea change in our attitude toward the search for extraterrestrial life. Before, looking for aliens was a disreputable fantasy. After, it became a valid scientific pursuit, especially at NASA. One key shift in perspective was the realization that we really did not know how to identify life when we saw it. Programs were quickly put in place to study extremophile bacteria and to upgrade NASA's experiments to detect life under extreme conditions. This led to the exciting development of post-Viking landers and rovers, and eventually to the current *Curiosity* rover, with its sophisticated chemistry laboratory. The ALH84001 meteorite has fostered quite a legacy of discovery, including the search for exoplanets beyond our solar system and the identification of liquid water under the surfaces of various moons of Jupiter and Saturn. One day, this research may reveal fossilized life forms—or, for that matter, still-living life forms—elsewhere in the universe.

90

Sojourner

**Robotic exploration
of Mars begins**

1997

A mobile camera on wheels, the *Sojourner* rover landed on Mars on July 4, 1997 on the back of the *Pathfinder* lander, in a region called Ares Vallis. It was the first such rover ever to touch down on another planet. For the next eighty-three days, under the control of engineers back on Earth, the twenty-five-pound robotic surface explorer roamed a modest 330 feet and returned some 550 images of the Martian surface. Meanwhile, the nearby *Pathfinder* lander transmitted more than 16,000 images.

The rover had three cameras: two monochrome cameras in front and a color camera in the rear. Both front cameras had a CCD imaging array measuring 484 pixels high by 768 pixels wide. Each camera weighed about one-and-a-half ounces and had a four-millimeter lens—not much larger than the diameter of a typical smartphone lens today. *Sojourner* sent its images back to the *Pathfinder*

lander, which had been christened the Carl Sagan Memorial Station after it touched down. Then the images were telemetered to Earth. A total of 287 megabytes of data were returned. With a resolution that was only slightly worse than "retinal" (the resolution of the human eye), the photographs were dramatic and history making, showing the planet from an entirely new perspective. And they were scientifically important, too, because they revealed that the climate on Mars was once warmer and wetter than it is now.

Groundbreaking though *Sojourner* was, the fundamental technology behind it was decidedly old-school. The Soviet *Lunokhod* lunar rovers had roamed the surface of the Moon back in the early 1970s. But there was a critical new challenge in operating a rover on Mars: While the *Lunokhod* rovers were controlled in real time by operators back on Earth, this approach would not work on Mars, where the radio delay could amount to twenty minutes each way. A lot of bad things can happen to a rover left to its own devices for forty minutes! And so *Sojourner* was programmed to be semiautonomous, carrying out some elements of its scientific experiments without needing constant input from its handlers on Earth. In that way, *Sojourner* isn't only a milestone in space exploration, but a landmark in robotics, too—an incredible machine capable of operating like an electronic geologist: surveying, mapping, and chemically testing a planetary surface millions of miles from the nearest human.

◀ Close-up of gyroscope motor
and housing halves

91

Gravity Probe B

Testing general relativity

2004

Gravity Probe B (GP-B) was a satellite mission launched by NASA on April 20, 2004, to test two unverified but important predictions made by Albert Einstein's theory of general relativity: the geodetic effect (the idea that space itself is elastic, and can stretch to absorb a particle's energy generated by its spin) and frame dragging (in which an object spinning in space will drag some of space-time along with it). They did this by measuring, very precisely, tiny changes in the direction of spin of four gyroscopes contained inside a satellite orbiting four hundred miles above Earth and crossing directly over the poles. Within their housings, the gyroscopes did not come into contact with any part of the surrounding satellite, meaning each of the spheres was essentially its own independent satellite orbiting Earth. They were monitored by sensors to determine their precise orientation in space.

According to Einstein's theory of general relativity, the spin axis of the gyroscope should shift by a very small angle over the course of the thousands of orbits that GP-B would take around Earth. But to convincingly measure this minute angular change, a near-perfect gyroscope had to be built. After years of work and the invention of new technologies, the result was a one-and-a-half-inch sphere of pure fused quartz, polished to within a few atomic layers of being perfectly smooth. According to Guinness World Records, these are the roundest objects ever made by humans. If a

GP-B gyroscope were enlarged to the size of Earth, its tallest mountain or deepest ocean trench would be only eight feet high (or deep)! Only neutron stars are more spherical. Another amazing thing about this gyro technology is that the spheres spin at four thousand revolutions per minute, but if left to themselves in the low-friction vacuum of space, it would take about fifteen thousand years for them to stop spinning.

By 2011, the frame-dragging effect was measured to be 37.2±7.2 milliarcseconds, and had been confirmed to within 5 percent of the expected result of 39.2 milliarcseconds. The geodetic effect, measured to be 6,602±18 milliarcseconds, was confirmed to better than 0.06 percent of the predicted value of 6,606 milliarcseconds. The GP-B remains one of the most precise tests of these important relativistic effects.

92

LIDAR

**Automated docking maneuvers
removing humans from the equation**

2007

I n the 1960s, Soviet cosmonauts were not allowed to exercise much autonomy while flying their spaceships; most flight control was performed by ground-based experts. Consequently, their space program began to aggressively pursue automated rendezvous and docking, which led to the successful docking of the two unmanned *Kosmos 186* and *188* spacecraft on October 30, 1967.

In contrast, beginning with the 1966 *Gemini 8* mission flown by Neil Armstrong and David Scott, the US space program depended upon the astronaut-intensive approach of fully manual docking. In fact, one of the Gemini program's major goals was to perfect these manual techniques for the Apollo program and the eventual docking of the lunar excursion module with the command module while in lunar orbit. This manual docking technique was later extended to the space shuttle and the International Space Station (ISS). But manual methods had at least one major drawback: They simply weren't an option for remote satellite servicing or space station resupply.

Decades later, light detection and ranging (LIDAR; also known as LADAR) would help NASA finally make the move to automatic. LIDAR works a lot like radar, detecting shape and distance of objects by sending out pulses of laser light (rather than radar's radio waves) and registering what bounces back. And it was used to dramatic

effect during the Department of Defense's Orbital Express mission in April 2007. Orbital Express consisted of two spacecraft that were able to separate and reattach under autonomous control using the Advanced Video Guidance Sensor (AVGS)—a kind of LIDAR system in which a laser illuminates a target to take its picture and, by using fixed retroreflectors on the target, figures the approach position and speed of the docking vehicle. It was the first rendezvous and docking without human aid in the history of the US Space Program.

But perhaps the greatest mark LIDAR has left on the world may be in the field of eye care. Autonomous Technologies applied the laser tracking technology it helped develop for NASA to tracking the eye during surgery, with the new operating technique going public in 1998 as LADARVision CustomCornea. Current LADARVision tracking systems make over four thousand measurements every second and can follow eye motion precisely while at the same time reshaping the cornea. Ironically, it wasn't until 2007 that NASA allowed astronauts to use laser surgery to correct their eyesight to the 20/20 vision required for spaceflight.

◀ Orbital Express, consisting
of two satellites

93

The Large Hadron Collider

The most complex machine ever built

2008

The universe is made of matter, and any proper understanding of how our universe is composed must rely on the most detailed picture of the nature of matter that we can put together. This is particularly important in studies of stellar evolution and the earliest moments in the origin of our universe through the Big Bang. In high-energy physics, the detailed structure of atomic and subatomic matter is laid bare using enormous instruments called *accelerators*.

When it comes to describing these tools of high-energy physics, you run out of superlatives quickly. The reason for this is that to see and measure very small particles that come and go inside of atoms, you have to use accelerators, colloquially called *atom smashers*, that concentrate lots of energy into a small amount of space. Thanks to Albert Einstein's famous formula $E=mc^2$, if new particles exist beyond the familiar electrons, protons, and neutrons, you can create them by colliding protons together at very high energy. Doing this takes very complex instruments indeed. First the colliding particles have to be accelerated to nearly the speed of light; then they have to be focused into a small beam to increase the number of collisions that will occur. From a detailed analysis of

the collision process, new kinds of particles can be created and studied in order to confirm or disprove specific theories of matter.

This is what the Large Hadron Collider (LHC) is able to do. It's the largest scientific instrument ever made—in fact, it's the largest machine on Earth, period. The LHC was built by an international consortium of physics labs and universities between 1998 and 2008. The experiments run by the LHC aim to breach an important energy level beyond which physicists theorize that a "new physics" might emerge—that is, phenomena that behave according to principles of physics different from those we're familiar with. Drawing on the experience of several generations of accelerator designers and the engineering of dozens of lower-energy laboratory machines, including Fermilab in the United States, the LHC—a massive seventeen-mile-long ring made of roughly nine thousand superconducting focusing magnets—was built more than three hundred feet underground near Geneva, Switzerland. But that was only the beginning. The ring would require one hundred megawatts of electricity from the local power grid to run, around a hundred tons of liquid helium to keep cool, and 1,800 miles of cables to carry power and data.

When operating, protons are circulated in opposite directions at 99.9999 percent of the speed of light. The beams are brought together at several points along the circumference, where giant detectors the size of small houses are ready to study the collisions. Millions of collisions occur every second, producing sprays of particles that penetrate deep into the detectors, where digital electronics and sensors record their properties in minute detail. The amount of data flowing from these collisions is so colossal—more than twenty-five thousand terabytes a year—that it has to be farmed out in near real-time to hundreds of supercomputing centers around the world for processing.

In 2012, the LHC made the historic discovery of the Higgs boson: the missing particle in a theory physicists call the standard model. Before it shut down for upgrades in late 2018, the LHC was testing the principles of the standard model at enormous energies of thirteen trillion volts (13 TeV) to see where the current theory of matter and forces breaks down, but no "new physics" has been found—yet.

94

The Kepler Space Telescope

World's largest digital camera in space

2009

◀ Kepler focal plane array

Imagine looking at a porch light on a warm summer evening as a large moth flies around it. As the moth crosses your sight line, it blocks some of the lamplight and the light fades just a bit. This basic idea has been used since the late 1990s as a means to detect planets orbiting other stars. By 2008, astronomers had detected more than 250 exoplanets as they transited their stars and eclipsed some of their light.

But the technology for measuring faint brightness shifts in stars was advancing considerably. A team of astronomers led by William Borucki at the Ames Research Center in California developed a new detection strategy: Use digital imaging technology to measure the brightness of thousands of stars at the same time. A single star image, with the right-sized telescope, would occupy a few pixels in the camera's field of view. If you measure the brightness of the star electronically, you can make frequent photographs of the same star field and capture the light changes of hundreds of thousands of stars every few minutes. This would reveal which stars were being orbited by objects.

The Kepler mission, launched on March 7, 2009 (and decommissioned in 2018), contained a fifty-five-inch telescope with a

twelve-degree field of view, housed in a spacecraft built by Ball Aerospace Corporation, that was kept pointed at the same star field in the constellation Cygnus. It did this hour after hour, day after day, by using precise reaction wheel gyroscopes. At the focus of the telescope was an advanced digital camera—the largest-format camera ever placed into space. It contained forty-two CCD imagers, each with a 2,200-by-1,024-pixel format totaling 95 megapixels. Through the telescope, this array could monitor more than 150,000 stars within a region of sky about 150 times the area of the full Moon.

The huge amount of data the camera generated every six seconds was too large to store in its entirety onboard the spacecraft, so the information to be kept (roughly 5 percent of the pixels) was chosen strategically and transmitted once a month. Nevertheless, hundreds of measurements could be made of each star every day, with brightness changes of as little as thirty parts per million for stars as faint as the twelfth magnitude. By comparison, the faintest stars you can see with the naked eye are about sixth magnitude, which means that these stars can be three hundred times fainter than what the unaided human eye can see in the sky. Kepler's advanced camera, with its

▲
Kepler's full field of view
from 42 CCD imagers

▲
Artistic rendering
of Kepler 186-f

very sensitive light-detection ability, allowed planets as small as Earth to be detected as they dimmed the light from their parent star.

In 2014, Kepler made a groundbreaking discovery: Kepler-186f, the first Earth-sized planet ever discovered in another solar system that has been confirmed to be in the habitable zone, the crucial range of distances from a star at which liquid water, widely considered a key ingredient for life, might collect on a planet's surface.

By 2018, more than 2,600 exoplanets had been discovered and confirmed by Kepler, and nearly 3,000 more had been detected but were yet to be confirmed. In our age-old quest to determine just how special our planet and the life on it truly is—if at all—the Kepler telescope has played a crucial role: it has opened our eyes to how many more planets are really out there that could potentially support life. Based on the number of exoplanets Kepler found that were both the size of Earth and located in their habitable zones, scientists have extrapolated that there could be billions of other planets in habitable zones and therefore, at least in theory, capable of sustaining life … at least the organic kind that we know about!

95

Curiosity Rover

**An astonishing robotic
space explorer**

2012

A self-portrait of *Curiosity* ▶

Every space scientist and astronomer knows exactly where they were on August 6, 2012, at 05:17 UTC, when the one-ton *Curiosity* rover, known more formally as the Mars Science Laboratory (MSL), made its harrowing descent to the surface of Mars. It wasn't the first rover on Mars—three came before—but it's the most advanced to date, and, as it approached the surface of the planet, it carried humanity's hope of discovering things about Mars that would fundamentally change the way we view it. It would not disappoint.

More than three million viewers watched the landing via video feeds streamed over the internet from the Jet Propulsion Laboratory in California. The landing came to be dubbed "seven minutes of terror," as scientists watched anxiously to see whether a precisely timed sequence of moves had unfolded successfully. Tension was heightened because these events all happened with a frustrating telemetry

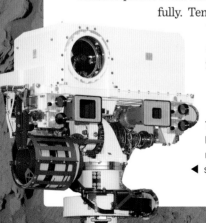

The remote sensing mast on the Mars Science Laboratory mission's rover, *Curiosity*, shows seven of ◀ seventeen cameras.

delay of fourteen minutes due to Mars's distance from Earth. But all went according to plan: The descent capsule's landing rockets fired, the heat shield came off, the parachutes unfurled, and the descent stage hovered at an altitude of around sixty feet before lowering the MSL safely to the ground in Gale Crater.

The *Curiosity* rover is the size of a small SUV, an articulated mobile laboratory and surveying instrument on steroids—the most recent in a long line of missions to Mars stretching back to the *Viking I* landing in 1976. It can photograph the landscape at extremely high resolution. It can drill rocks and process the samples in a chemical laboratory to identify exact mineral and compound types. With its radiation and gas sensors it can measure environmental levels that are important for future astronauts to know about. To run all of these instruments and transmit data to the Mars *Orbiter* satellite for relay to Earth, it uses a 110-watt radioisotope thermoelectric generator (RTG) powered by the decay of radioactive plutonium. It was designed to survive for only 700 days. Yet by the end of 2018, after 2,300 days of active operation, *Curiosity* was still functioning. It had traveled nearly twelve miles across the crater floor and had already spent four years exploring the foothills of the central mountain, called Mount Sharp.

Nevertheless, it has a long way to go before it surpasses the 2004 *Opportunity* rover's twelve-year life span. (*Opportunity* was predicted to stop working after only three months.)

The *Curiosity* rover's Mastcam is one of its most recognizable features, located on a six-foot-tall boom and designed to take panorama photos of the landscape. It can take and store up to several hours of HD video or five thousand high-resolution color photos, comparable to those taken by a two-megapixel smartphone camera. Aside from returning amazing images of a wide range of Martian geological formations and landscapes, *Curiosity* has found evidence of an ancient stream bed where water once flowed for thousands of years. It has also determined that the radiation levels on the surface are no worse than what astronauts experience on the International Space Station. The rover's chemistry laboratory has detected elements such as sulfur, nitrogen, and phosphorus, all of which are crucial for life, as well as clays that indicate that large bodies of standing water existed in Gale Crater at some time in the past. *Curiosity* has also confirmed the seasonal presence of methane gas. Whether this gas is being produced by organic or inorganic sources remains a tantalizing question for future rovers to explore.

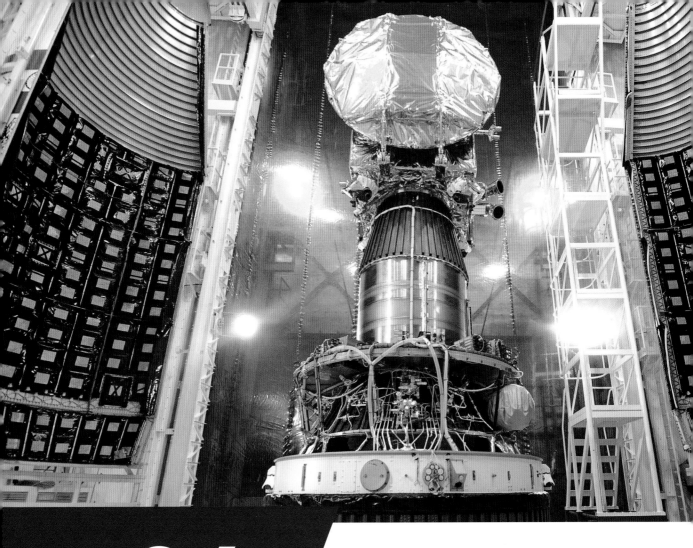

96

Mangalyaan— the Mars Orbiter Mission

India joins Club Mars—at a discount

November 2013

On November 5, 2013, the Indian Space Research Organization (ISRO) launched the Mars Orbiter, known as *Mangalyaan*. Outwardly, there is nothing remarkable about such an event. Since the 1960s, no fewer than twelve nations, including India, have embarked on a satellite launch program. But what makes this orbiter, and its final arrival at Mars on September 24, 2014, so significant is in fact twofold: First, ISRO became only the fourth space program to successfully reach Mars, joining the United States (NASA), Europe (ESA), and Russia (Roscosmos), as well as the space program of the former Soviet Union in this very elite club. Moreover, two-thirds of the forty-eight spacecraft sent to Mars have failed, making *Mangalyaan*'s success a technically remarkable achievement.

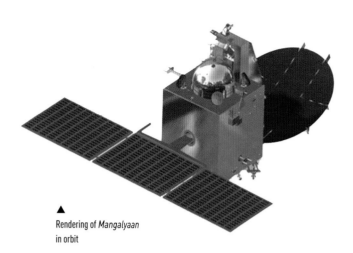

◀ *Mangalyaan* preparing
for launch

▲
Rendering of *Mangalyaan*
in orbit

The second reason *Mangalyaan* is so noteworthy is the astonishingly modest cost of the mission. The spacecraft itself cost only $25 million, making it the least-expensive of its kind to successfully reach Mars. The savings was attributed to the use of only a very few (a mere five) instruments, plus a modular design borrowing components that were "off the shelf" in the Indian space agency. There was no extensive testing of the spacecraft, and from start to finish it was assembled and made flight-ready in only fifteen months. Needless to say, the launch and successful arrival of *Mangalyaan* at Mars was received with tremendous celebration by the Indian public, and its likeness can now be found on the back of the two-thousand-rupee note. A photograph of the Indian Mission Control Center after the successful orbit insertion showed a number of female scientists and engineers in colorful saris celebrating, which was hailed as a major empowering event for India's women in their struggle for equality.

The fueled spacecraft weighed about 3,000 pounds (1,400 kg), of which the scientific payload accounted for only 33 pounds (15 kg).

Its three solar panels, with a combined surface area of 240 square feet (22 m²), generated a modest 840 watts of power. The electricity was stored in a 36 amp-hour lithium-ion battery system. Its 7-foot (2.2-meter) dish provided telemetry contact with Earth through India's Deep Space Network of ground stations.

Mangalyaan's scientific mission was to take full-color images of the entire disk of Mars from a highly elliptical orbit, making it the only Mars-orbiting spacecraft able to provide such full-Mars images on a regular basis. Its simple Lyman Alpha Photometer detects the emission from escaping hydrogen and deuterium gas from the planet's atmosphere, allowing scientists to determine the amount of water loss by atmospheric evaporation. It will partner with NASA's much more expensive and scientifically-complex MAVEN spacecraft (which costs $671 million) in studying the Martian atmosphere over long periods of time. And its Thermal Infrared Imaging Spectrometer will map the surface temperature and composition of Mars and combine its data with the Mars Color Camera to obtain high-resolution mineralogy maps of the Martian surface and monitor for dust storms and other meteorological events. Not a bad bargain for such an inexpensive spacecraft. With governments constantly threatening to scale back and defund their space programs, *Mangalyaan* may show a way forward for programs to try to make the most of the funding they have.

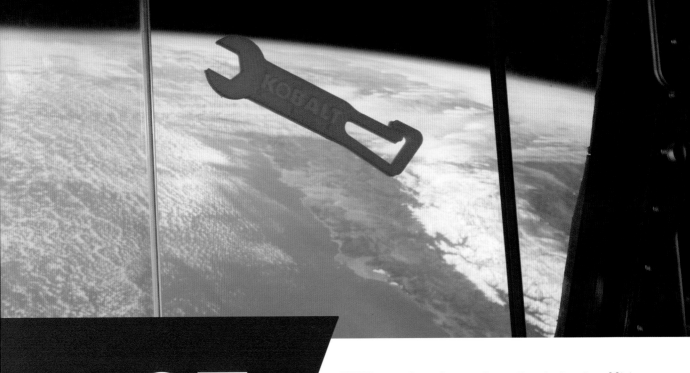

97

A 3-D Printed Ratchet Wrench

Print what you need, when you need it, in space

2014

The technical term for 3-D printing is *additive manufacturing,* because the process involves addition rather than subtraction: Instead of putting a piece of metal on a lathe and fashioning an object by removing what you don't need, why not form the item by extruding metal and building it up, layer by layer? The technology behind 3-D printing was first developed in 1981 for creating plastic parts; since then, it's fallen dramatically in price—by 2018, a few hundred dollars could buy a computer-driven system of considerable versatility—inspiring new commercial applications as well as uses for educators and hobbyists.

NASA became interested in 3-D printing as an onsite manufacturing technique. It takes a lot of time and resources to launch objects made on Earth up into space, and using 3-D printers would allow missions to skip that step entirely. Instead, astronauts could fabricate replacement parts and tools on the International Space Station itself by merely uploading a suitable print file and having the ISS printer fabricate the part. This had the potential to be a huge improvement on having to stock the ISS with replacement parts—not

An example of the ratchet wrench, which was NASA's first tool made with 3-D printing in space.

SpaceX is a leader in the 3-D printing of rocket engine components, such as those for their Merlin engines.

only would it cost less, but it would also save valuable storage space.

In 2014, NASA tested its theory by printing its first tool aboard the ISS using a file uploaded from Earth: a plastic ratchet wrench. Designed by Noah Paul-Gin of Made In Space, Inc.—a company hired by NASA to design, build, and operate the printer—the wrench, five inches long by one-and-a-half inches wide, took less than a week to design and approve for printing; the actual printing of the object itself took a mere four hours. The process is a game changer, capable of turning the process of restocking the space station, which can take many months, into a simple print job that can be completed far more quickly, and only for those replacement items you actually need.

Thus far, additive manufacturing of nonplastic aerospace components is still in its infancy, but the momentum behind this technology is growing. In 2013, the rocket chambers for SpaceX's SuperDraco engines were 3-D printed—albeit on Earth—as were the main oxidizer valves for its Merlin 1D engines in 2014. In the aeronautical industry, Aerojet Rocketdyne printed the copper-alloy thrust chamber for its RL10 rocket engine in 2017. The expectation is that in the next decade, progressively larger segments of rocket engine and spacecraft systems will be printed. Some scientists even envision the printing of lunar and Martian colony dwellings. Freed from having to depend on Earth for supplies, long-range, long-term deep-space colonization has taken a leap forward into reality.

Both NASA and the ESA are developing ground tests in which buildings will be fabricated using 3-D printing technology. The idea is something like a mobile cement factory: Printing material would be created from local ingredients and combined with a binder, then extruded to form multistory buildings from a digital blueprint long before inhabitants arrived.

98

The LIGO Gravitational-Wave Interferometer

Finding ripples in space-time

2015

In 1915, Albert Einstein published his theory of general relativity, which described gravity as a distortion in the curvature of space-time. It wasn't much of a leap for him to also realize that changes in a gravitational field caused by accelerating masses would cause changes in the curvature of space, which would travel outward at the speed of light, leading to gravitational waves.

But his theory of gravitational waves would remain just that—a theory—until someone actually observed them. In the 1960s, Joseph Weber, a professor at the University of Maryland, built the first gravitational wave detector: multiple solid aluminum cylinders that each weighed more than a ton and were equipped with ultra-sensitive strain gauges. When a gravitational wave of the right frequency passed through the lab, it would

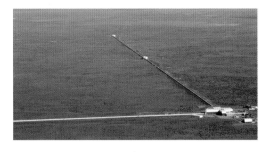

◀ LIGO in
Livingston, LA

▲
LIGO's counterpart in
Hanford, WA

cause one or more of the cylinders, or "bars," to change their dimensions, and therefore vibrate, which would be seen in the strain gauges. No confirmed events were ever detected in this way, but Weber's work got the physics community interested in looking for gravitational waves, leading to more elaborate designs for detectors, and ultimately far more precise ones.

One of the most sensitive ways to measure minute changes in distance is to use an interferometer, invented by the American physicist Albert Michelson in 1887. In his device, a single beam of light is sent to a partially reflective mirror called a *beam-splitter,* which reflects 50 percent of the light to a second mirror, placed at a right angle to the original beam of light. The other 50 percent of the light passes through to a third mirror. Both beams, or "arms," of light are then reflected back to the beam-splitter, where the combined light from the two different paths interferes to produce fringes, or patterns of dark and light areas. In a gravitational wave interferometer, the length of each arm would change in a specific way in response to the appearance of a gravitational wave, which would be seen in how the fringe patterns change over time.

In 1994, construction of the Laser Interferometer Gravitational-Wave Observatory (LIGO) began at the Hanford Site, a decommissioned nuclear production complex in Washington state, along with a twin system in Livingston, Louisiana (having two stations eliminates local sources of noise). It would be the world's largest observatory of its kind. Each system is made of two concrete tubes, each two and a half miles long, for lasers to travel down and be reflected by mirrors. An elaborate optical system can measure changes in the lengths of the lasers, or "arms," to $1/10,000$ of the diameter of a proton. This is equivalent to measuring the distance from Earth to the nearby star Alpha Centauri to an accuracy of one millimeter.

In 2015, not long after the observatory commenced operating but a century after Einstein had made his prediction, LIGO detected a gravitational wave event for the first time. Scientists believe it was the result of two massive black holes circling each other at half the speed of light, then merging into one, bending space-time and releasing gravitational waves in the process.

By the end of 2018, eleven gravitational wave events had been detected, resulting in a growing catalog of wave originators across the cosmos. Meanwhile, a careful study of the shapes of the waves in time and space has led to detailed models of the sources. The simplest explanation—one that precisely corresponds with the shape of the received pulses—is that most of these events were black holes merging with each other in binary systems a billion light-years from the Sun. Long after Einstein's brilliant insight, LIGO has confirmed a monumental theory of the way space works and given us access to a whole new way of seeing the cosmos.

99

The Tesla Roadster

Advertising enters the Space Age

February 2018

Why use a dummy payload made of cement when you can launch your own sports car? SpaceX, Elon Musk's aerospace company, has broken all kinds of barriers in what was once the government-dominated world of space travel, including the 2010 cargo spacecraft *Dragon*'s successful launch, orbit, and recovery—a first for a private company. Perhaps SpaceX's most notorious and groundbreaking act to date was its test of the *Falcon Heavy*, a heavy-payload vehicle, not as much for the spacecraft itself but for what it contained. As Musk tweeted out on December 1, 2017: "Payload will be my midnight cherry Tesla Roadster playing Space Oddity. Destination is Mars orbit. Will be in deep space for a billion years or so if it doesn't blow up on ascent." It was launched on February 6, 2018.

The Roadster is a landmark in our space history because it is the quintessential symbol of this new era of private space travel: an entrepreneur's own car company creating the payload for his space company's own rocket. NASA's funding has shrunk far below its '60s peak—its roughly $21 billion dollars today represent less than half of 1 percent of the federal budget, which is less than an eighth of its 1966 high-water mark—while the number of private space companies (and their respective budgets) continues to grow. It's clear that the future of space exploration increasingly depends upon private enterprise.

The Roadster's noteworthy for its ushering consumerism into space, too. It's an object that you could go out and buy yourself, and therefore is something like the first-ever advertisement in space—a major threshold crossed in the commercialization of the cosmos.

And arguably the most important point of all: The thought of a sports car speeding through space stirs our sense of wonder. And that's how the next great milestones in space exploration are made. Just as with President John F. Kennedy's famous declaration that an American would land on the Moon before the decade was out, a bold promise that captured the country's imagination and helped propel NASA toward its goal, a Roadster blazing through our solar system is outlandish enough that it expands the possibilities of what we think is within reach. As Musk later noted, "I love the thought of a car drifting apparently endlessly through space and perhaps being discovered by an alien race millions of years in the future." Note to Musk: So do many astronomers!

The last reported visual contact with the Roadster was by astronomers at the University of Arizona later in February 2018, by which time it had already traveled three million miles. As of this writing, it's more than two hundred million miles away.

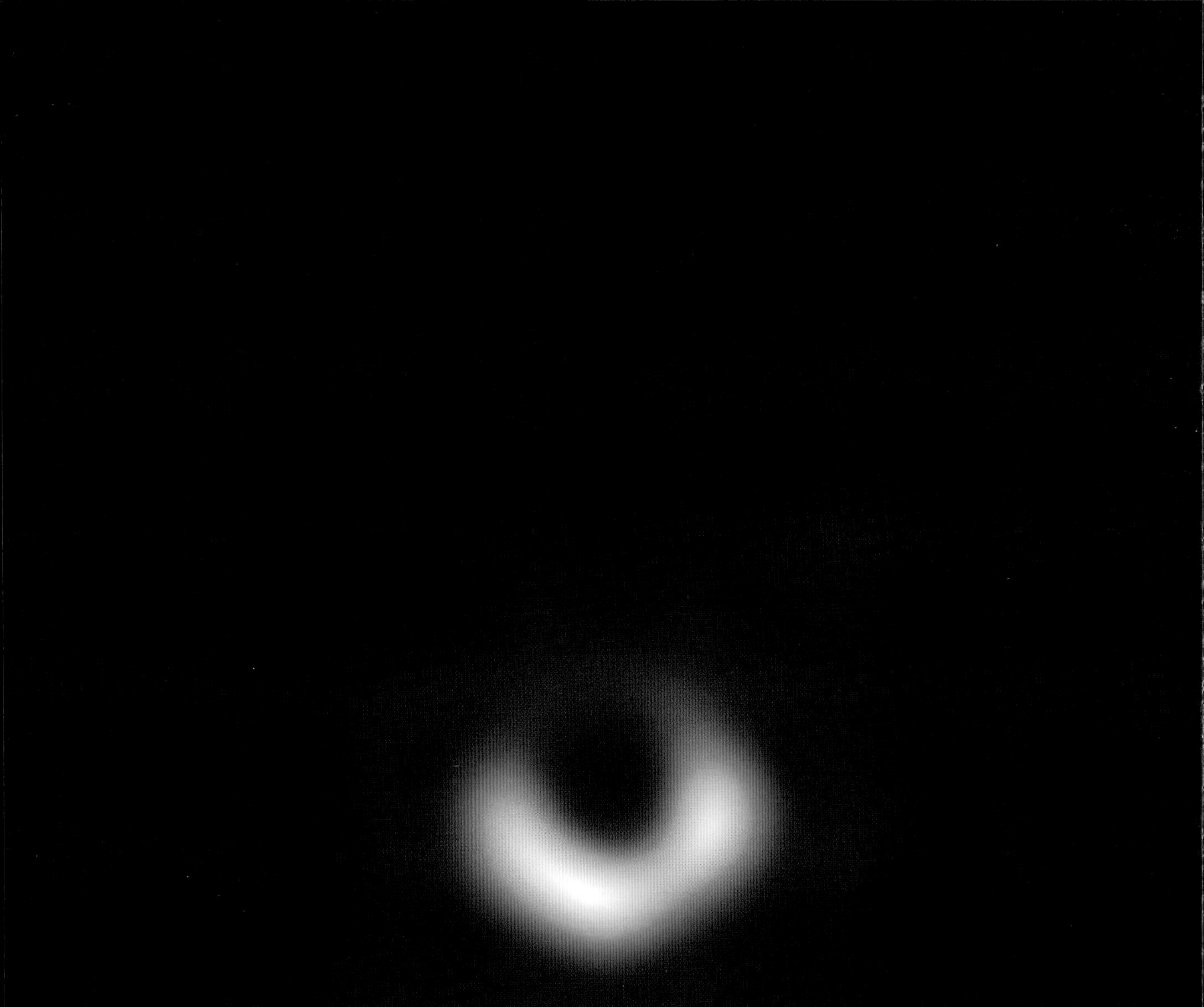

100

The Event Horizon Telescope

The first glimpse of a black hole

2019

◀ The synthesized image of the M87 supermassive black hole: The dark spot at the center is the shadow of the black hole projected against the infalling, luminous matter. The scale of this image is such that our entire solar system, all the way to Pluto, would fit inside the dark event-horizon zone. The surrounding matter in the accretion disk is orbiting clockwise at nearly the speed of light.

The field of radio interferometry, initially developed in 1946 using pairs of radio telescopes to synthesize high-resolution images, steadily expanded over time as technology improved. The signals received by each pair of telescopes had to be recorded, first using analog videotapes, then using high-speed digital recording on high-capacity computer hard drives. The time signals had to be tagged with their arrival times using atomic clocks, which steadily improved in their stability over the decades. Computer processing technology had to advance to the point where supercomputers could perform the trillions of computations each second that were necessary to efficiently correlate radio signals across hundreds of pairs of telescopes in an interferometer array. Finally, the radio receiver technology had to improve dramatically in order to allow signals at progressively higher frequencies to be detected with a minimum of noise. By 2018, all of these improvements had come together, allowing for the creation of a massive interferometer called the Event Horizon Telescope (EHT).

The EHT began as a network of eight radio telescopes, all located in the western hemisphere, that pair up to gaze at different parts of the skies. Operating at a wavelength of 1.3 millimeters, this "aperture synthesis" telescope has a resolution of twenty microarcseconds. Its first goal was to capture an image of the plasma near the event horizon of the supermassive black hole in the galaxy Messier 87 (M87), an object with an estimated mass nearly seven billion times that of our Sun. In order to do this, each telescope in the network collected roughly 350 terabytes of data every day. Specially designed supercomputers then sifted through petabytes of time-tagged data to identify the same radio wave-front arriving at all of the telescopes, and from this, an image of the plasma near the forty-billion-kilometer-long event horizon was synthesized. The result was announced at a worldwide press conference on April 10, 2019: the first-ever image of an actual black hole, large enough to swallow our entire solar system. Future studies will include following the month-to-month changes in the black hole's inflowing plasma and later turning the EHT toward planets being formed around distant stars.

◀ The current Event Horizon Telescope is a network of radio telescopes that are able to act as a single telescope spanning nearly the diameter of Earth.

Credit: EHT Collaboration

RESOURCES AND PHOTO CREDITS

The information in this book was drawn largely from my own expertise as well as a number of essential sources, including NASA, *Smithsonian* magazine, Space.com, and Britannica.com. Apart from visiting many of these objects in museums and collections around the world, you can learn more about them by taking your own online journey; I'd recommend these key sources as a launchpad for your travels! The sources below represent those that were specifically used for individual entries.

The photo credits below refer to the images on the book's interior, back cover, and poster.

1 The Blombos Ochre Drawing
"500,000-Year-Old Homo erectus Engraving Discovered," SciNews (December 4, 2014) • Bradshaw Foundation: bradshawfoundation.com • Chutel, Lynsey, "What the Oldest Drawing Found in South Africa Tells Us About Our Human Ancestors," Quartz Africa (September 16, 2018) • D'Errico, Francesco; Henshilwood, Christopher S.; Watts, Ian, "Engraved Ochres from the Middle Stone Age Levels at Blombos Cave, South Africa," *Journal of Human Evolution* 57(1): 27–47, July 2009 • Gabbatiss, Josh, "Oldest Drawing Ever Found Discovered in South African Cave, Archaeologists Say," *The Independent* (September 12, 2018) • St. Fleur, Nicholas, "Oldest Known Drawing by Human Hands Discovered in South African Cave," *The New York Times* (September 12, 2018)
PHOTO CREDIT: Image © Craig Foster. Courtesy of Professor Christopher Henshilwood.

2 The Abri Blanchard Bone Plaque
Cave Script Translation Project: cavescript.org • Feder, Kenneth L., *Encyclopedia of Dubious Archaeology*, 2010, Santa Barbara, CA: Greenwood
PHOTO CREDIT: Gift of Elaine F. Marshack, 2005. Courtesy of the Peabody Museum of Archaeology and Ethnology, Harvard University.

3 The Egyptian Star Clock
Bryner, Jeanna, "Ancient Egyptian Sundial Discovered at Valley of the Kings," LiveScience (March 20, 201 3)
PHOTO CREDIT: Wikipedia/Einsamer Schütze. Distributed under the CC BY-SA 3.0 license.

4 The Nebra Sky Disk
Haughton, Brian, "The Nebra Sky Disc—Ancient Map of the Stars," Ancient History Encyclopedia (May 10, 2011)
PHOTO CREDIT: Wikipedia/Anagoria. Distributed under the CC BY-SA 3.0 license.

5 The Venus Tablet of Ammisaduqa
Khan Academy: khanacademy.org • Novakovic, B., "Senenmut: An Ancient Egyptian Astronomer," *Publications of the Astronomical Observatory of Belgrade* 85: 19–23, 2008 • Radeska, Tijana, "The Royal Library of Ashurbanipal Had Over 30,000 Clay Tablets, Among Them Is the Original 'Epic of Gilgamesh,'" *The Vintage News* (November 30 2016)
PHOTO CREDIT: Wikipedia/Fæ. Distributed under the CC BY-SA 3.0 license.

6 The Star Charts of Senenmut
Ancient Egypt Online: ancient-egypt-online.com • Belmonte, Juan Antonio; Shaltout, Mosalam, "The Astronomical Ceiling of Senenmut: A Dream of Mystery and Imagination," European Society for Astronomy in Culture, 2005 • Belmonte, Juan Antonio; Shaltout, Mosalam, In Search of Cosmic Order: Selected Essays on Egyptian Archaeoastronomy, Supreme Council of Antiquities Press, 2009 • Berio, Alessandro, "The Celestial River: Identifying the Ancient Egyptian Constellations," *Sino-Platonic Papers* 253, 2014 • The Earth Chronicles of Life: earth-chronicles.com • Mills, Thomas O., "Star Maps and the Secrets of Senenmut: Astronomical Ceilings and the Hopi Vision of Earth," Ancient Origins (November 18, 2016)
PHOTO CREDIT: Courtesy of the Rogers Fund, 1948.

7 The Merkhet
Ancient Egyptian Astronomy Database: aea.physics.mcmaster.ca • Ancient Pages: ancientpages.com • Louvre Museum: louvre.fr • The Metropolitan Museum of Art: metmuseum.org • Quantum Gaze: quantumgaze.com • WiseGeek: wisegeek.com
PHOTO CREDIT: Wikipedia/Rama. Distributed under the CC BY-SA 3.0 France license.

8 The Nimrud Lens
The British Museum: britishmuseum.org • Holloway, April, "Is the Assyrian Nimrud Lens the Oldest Telescope in the World?," Ancient Origins (February 24, 2014) • Whitehouse, David, "World's Oldest Telescope?," *BBC News* (July 1, 1999)
PHOTO CREDIT: Courtesy of the British Museum.

9 The Greek Armillary Sphere
MacTutor History of Mathematics Archive: history.mcs.st-and.ac.uk • The Metropolitan Museum of Art
PHOTO CREDITS: Peter Horree/Alamy Stock Photo. Inset: Wikipedia/-Merce-. Distributed under the CC BY-SA 3.0 license.

10 The Diopter
Kotsanas Museum of Ancient Greek Technology: kotsanas.com • Roman Aqueducts: romanaqueducts.info
PHOTO CREDITS: Drawing by Jack Dunnington (reconstruction of Heron's dioptra). Inset: Reconstruction of a Dioptra by Jens Kleb, Erfurt, Germany, 2014.

11 Antikythera Mechanism
Antikythera Mechanism: antikytheramechanism.com • European Physical Society: epsnews.eu • National Archaeology Museum: namuseum.gr/en • Trimmis, K. P., "The Forgotten Pioneer: Valerios Stais and His Research in Kythera, Antikythera, and Thessaly," *Bulletin of the History of Archaeology* 26(1), 2016
PHOTO CREDITS: Wikipedia/Tilemahos Efthimiadis. Distributed under the CC BY-SA 2.0 license. *Inset:* Have Camera Will Travel | Europe/Alamy Stock Photo.

12 Hipparchus's Star Atlas
Burnham, Robert, "Hipparchus's Sky Catalog Found," *Astronomy* (January 13, 2005)
PHOTO CREDITS: adam eastland/Alamy Stock Photo. *Inset:* Courtesy of Architectura database (architectura.cesr.univ-tours.fr).

13 The Astrolabe
Consortium for History of Science, Technology, and Medicine: chstm.org • The Mariners' Museum and Park: exploration.marinersmuseum.org
PHOTO CREDIT: Wikipedia/Sage Ross. Distributed under the CC BY-SA 3.0 license.

14 The Dunhuang Star Atlas
Bonnet-Bidaud, Jean-Marc; Praderie, Françoise; Whitfield, Susan, "The Dunhuang Chinese Sky: A Comprehensive Study of the Oldest Known Star Atlas," *Journal of History and Heritage* 12(1): 39–59, 2009 • The Iris: blogs.getty.edu/iris • Khan Academy
PHOTO CREDIT: Public Domain.

15 Al-Khwārizmī's Algebra Textbook
Today I Found Out: todayifoundout.com • World Digital Library: wdl.org
PHOTO CREDITS: Public Domain (both images).

16 The *Dresden Codex*
Vance, Erik, "Have We Been Misreading a Crucial Maya Codex for Centuries?," *National Geographic* (August 23, 2016)
PHOTO CREDITS: Wikipedia/Linear77. Distributed under the CC BY-SA 3.0 license.

17 The Chaco Canyon Sun Dagger
Exploratorium: exploratorium.edu • Imaging Research Center: irc.umbc.edu
PHOTO CREDIT: Charles Walker Collection/Alamy Stock Photo.

18 Giovanni de' Dondi's Astrarium
Poulle, E., "Book Review: The De' Dondi Astrarium," *Journal for the History of Astronomy* 20, 1989
PHOTO CREDITS: Wikipedia/Pippa Luigi/ Museo nazionale della scienza e della tecnologia Leonardo da Vinci, Milano. Distributed under the CC BY-SA 4.0 license.

19 The Big Horn Medicine Wheel
Hill, Pat, "The Mystery of the Big Horn Medicine Wheel," *Montana Pioneer*, May 2012 • Stanford Solar Center: solar-center.stanford.edu • US Department of Agriculture Forest Service: fs.usda.gov
PHOTO CREDITS: Photo Courtesy of Richard Collier, Wyoming State Historic Preservation Office. *Inset:* Drawing by Jack Dunnington (after an image from thescientificodyssey.typepad.com).

20 The Ensisheim Stone
Garber, Megan, "Thunderstone: What People Thought About Meteorites Before Modern Astronomy," *The Atlantic* (February 15, 2013) • Horejsi, Martin, "Ensisheim! The King of Meteorites," Meteorite Times Magazine (November 1, 2010) • Marvin, U. B., "The Meteorite of Ensisheim—1492 to 1992," *Meteoritics* 27, 28–72, 1992 • Rowland, I. D., "A Contemporary Account of the Ensisheim Meteorite, 1492," *Meteoritics* 25(1): 19, 1990 • Science Photo Library: sciencephoto.com
PHOTO CREDITS: Wikipedia/Daderot. Distributed under the CC BY-SA 2.0 license. *Inset:* Public Domain.

21 *De Revolutionibus*
DeMarco, Peter, "Book Quest Took Him Around the Globe," *Boston Globe* (April 13, 2004) • Wilford, John Noble, "Chasing Copernicus," *The New York Times* (July 18, 2004) • University of Glasgow Special Collections: special.lib.gla.ac.uk
PHOTO CREDIT: Public Domain.

22 Tycho's Mural Quadrant
Horrocks, Jeremiah, "The Transit of Venus and the 'New Astronomy' in Early Seventeenth-Century England," *Quarterly Journal of the Royal Astronomical Society* 31: 333, 1990 • National Center for Atmospheric Research, High Altitude Observatory: www2.hao.ucar.edu
PHOTO CREDIT: Public Domain.

23 Galileo's Telescope
Kestenholz, Daniel, "The Focal Length Closest to the Human Eye," Photography Daily Theme (September 29, 2012) • Universe Today: universetoday.com
PHOTO CREDITS: Getty Images/Leemage/Contributor. *Inset:* Public Domain.

24 The Slide Rule
History-Computer: history-computer.com • Just Collecting, Space Memorabilia: just-collecting.com/space-memorabilia • Space Flown Artifacts: spaceflownartifacts.com
PHOTO CREDITS: NASA. *Inset:* Wikipedia/Joe Haupt.

25 The Eyepiece Micrometer
Kaler, James B., Professor Emeritus of Astronomy, University of Illinois: stars.astro.illinois.edu • Mayer, Christian, "Directory of All Hitherto Discovered Doubled Stars," 1781 (accessed at spider.seds.org) • Niemela, V., "A Short History and Other Stories of Binary Stars," IX Latin American Regional IAU Meeting: Focal Points in Latin American Astronomy, Tonantzintla, Mexico (November 9–13, 1998)
PHOTO CREDITS: Public Domain (both images).

26 The Clock Drive
Biography: biography.com
PHOTO CREDITS: A. Duro/ESO. *Inset:* Courtesy of Judy Cleland Bergen.

27 The Meridian Circle
Nielsen, Axel V., "Ole Rømer and his Meridian Circle," *Vistas in Astronomy* 10 (Arthur Beer, ed.), Pergamon Press
PHOTO CREDIT: Wikipedia/Tsui. *Inset:* Distributed under the CC BY-SA 3.0 license.

28 The Skidi Pawnee Star Chart
Ancient Pages • Gustavus Adolphus College Physics Department: physics.gac.edu • Pasztor, Emilia; Rosland, Curt, "An Interpretation of the Nebra Disc," *Antiquity* 81(312): 267–78, 2007 • Pawnee Nation of Oklahoma: pawneenation.org
PHOTO CREDIT: Heritage Image Partnership Ltd./Alamy Stock Photo.

29 Smoked-Glass Sun Viewing
Historic New England: historicnewengland.org
PHOTO CREDITS: Public Domain. *Inset:* Wikipedia/Eclipse Glasses. Distributed under the CC BY-SA 3.0 license.

30 The Gyroscope
photo credits: NASA. Inset: Distributed under the GNU Free Documentation License

31 The Electric Battery
The British Museum • Deffner, Sebastian; Ibrahim, Muhammed, "Static Electricity's Tiny Sparks," The Conversation (January 6, 2017; accessed at phys.org) • Frank, Harvey; Halpert, Gerald; Surampudi, Subbarao, "Batteries and Fuel Cells in Space," Interface, The Electrochemical Society, Fall 1999 • Hubble Space Telescope: spacetelescope.org • Meyer, Michal, "Leyden Jar Battery," Distillations, Science History Institute (May 18, 2012)
PHOTO CREDITS: Van Leest Antiques, Utrecht (Leyden Jar). Wikipedia/GuidoB (Volta Battery); distributed under the CC BY-SA 3.0 license.

32 Pilâtre de Rozier and d'Arlandes's Balloon
Century of Flight: century-of-flight.net • CERN: cern.ch • Linda Hall Library: lindahall.org • Millikan, R. A.; Cameron, G. H., "The Origin of Cosmic Rays," *Physical Review* 32(533), 1928 • Pfotzer, G., "History of the Use of Balloons in Scientific Experiments," *Space Science Reviews* 13(2): 199–242, 1972 • This Day In Aviation: thisdayinaviation.com
PHOTO CREDIT: Public Domain.

33 William Herschel's Forty-Foot Telescope
Earth & Sky: earthsky.org • Herschel, William, "Catalogue of One Thousand New Nebulae and Clusters of Stars," *Philosophical Transactions of the Royal Society*, 1786 (accessed at royalsocietypublishing.org) • Peterson, Caroline Collins, "Meet William Herschel: Astronomer and Musician," ThoughtCo (July 3, 2019) • Science History Institute • Science Museum Group: sciencemuseum.org.uk
PHOTO CREDIT: Public Domain. Courtesy of The University of Chicago Library.

34   The Spectroscope
Fraunhofer-Gesellschaft: fraunhofer.de/en • Hiroshi Sugimoto: sugimotohiroshi.com • Lord Rayleigh, "Newton as an Experimenter," *Proceedings of the Royal Society of London* 131(864): 224–230, 1943 • New World Encyclopedia: newworldencyclopedia.org
PHOTO CREDITS: Public Domain (both images).

35   The Daguerreotype Camera
APS News, American Physical Society: aps.org • Hastings Historical Society: hastingshistoricalsociety.blogspot.com • Lights in the Dark by Jason Major: lightsinthedark.com • Taylor, Alan, "The Gift of the Daguerreotype," *The Atlantic* (August 19, 2015) • Trombino, Don, "Dr John William Draper," *Journal of the British Astronomical Association* 90: 565–571, 1980
PHOTO CREDITS: Public Domain (both images).

36   The Solar Panel
Espinoza, Javier, "Private Players Plug In to the Green Energy Revolution," *Financial Times* (November 28, 2018) • Love, Zen, "The First Solar-Powered Watch Was Far Ahead of Its Time," Gear Patrol (May 20, 2019) • PV Lighthouse: www2.pvlighthouse.com.au • Solar Cell Central: solarcellcentral.com
PHOTO CREDITS: NASA. *Inset:* Anthony Skelton.

37   The Leviathan of Parsonstown
Khan, Amina, "At Mt. Wilson, Scientists Celebrate 100th Birthday of the Telescope that Revealed the Universe," *Los Angeles Times* (November 1, 2017) • Messier Objects: messier-objects.com • Palomar Observatory: astro.caltech.edu
PHOTO CREDITS: Jared Enos. *Inset:* Public Domain.

38   Crookes Tube
Molecular Expressions: micro.magnet.fsu.edu • National High Magnetic Field Laboratory: nationalmaglab.org • North Arizona University Electron Microanalysis: www2.nau.edu/micro-analysis/wordpress • Sella, Andrea, "Aston's Mass Spectrograph," Chemistry World (July 3, 2014) • Thomson, Joseph John, "Rays of Positive Electricity," *Proceedings of the Royal Society* 89, 1913
PHOTO CREDITS: Wikipedia/D-Kuru. Distributed under the CC BY-SA 3.0 license. *Inset:* Public Domain.

39   The Triode Vacuum Tube
Electronics Notes: electronics-notes.com • Engineering and Technology History: ethw.org
PHOTO CREDIT: Wikipedia/Gregory F. Maxwell. Distributed under the GNU Free Documentation License.

40   The Ion Rocket Engine
Google Patents: patents.google.com
PHOTO CREDITS: NASA (both images).

41   The Hooker Telescope
Amazing Space: history.amazingspace.org • American Society of Mechanical Engineers: asme.org • Mount Wilson Observatory: mtwilson.edu • SpaceWatchtower: spacewatchtower.blogspot.com
PHOTO CREDIT: Wikipedia/Ken Spencer. Distributed under the CC BY-SA 3.0 license.

42   Robert Goddard's Rocket
PHOTO CREDITS: NASA. *Inset:* NASA.

43   The Van de Graaff Generator
Architectural Afterlife: architecturalafterlife.com • Lewis, Tanya, "Incredible Technology: How Atom Smashers Work," LiveScience (August 12, 2013) • Photographs of abandoned places, by Tom Kirsch: opacity.us
PHOTO CREDITS: AIP Emilio Segrè Visual Archives. *Inset:* Public Domain.

44   The Coronagraph
PHOTO CREDITS: ESO. *Inset:* NASA.

45   Jansky's Merry-Go-Round Radio Telescope
American Astronomical Society: aas.org • National Radio Astronomy Observatory: nrao.edu
PHOTO CREDIT: NRAO/AUI/NSF

46   The V-2 Rocket
Dean, James, "65 Years Ago, Cape Took Flight with Bumper 8," *Florida Today* (July 25, 2015) • Evans, Ben, "A Bumper Crop: The Cape's First Roar of Rocket Engines," AmericaSpace (June 24, 2012) • Messier, Doug, "Where the Space Age Really Began," Parabolic Arc (October 3, 2016) • This Day In Navigation
PHOTO CREDITS: Wikipedia/NASA/US Army. *Inset:* Wikipedia/Bairuilong.

47   ENIAC
Farrington, Gregory C., "ENIAC: The Birth of the Information Age," *Popular Science*, March 1996 • IBM: ibm.com • "When Computer Bugs Were Actual Insects," OpenMind (November 2, 2015)
PHOTO CREDIT: Public Domain. *Inset:* Wikipedia/IBM Italia. Distributed under the CC BY-SA 4.0 license.

48   Colossus Mark 2
Oak Ridge National Laboratory: ornl.gov • Stanford University Computer Science: cs.stanford.edu
PHOTO CREDIT: Wikipedia/Ibonzer. Distributed under the CC BY-SA 3.0 license.

49   The Radio Interferometer
Abshier, Jim, "Amateur Radio Astronomy: 400 MHz Interferometer," *Reflections of the University Lowbrow Astronomers*, March 2007 (accessed at umich.edu) • Bai, Xuening, "Radio Interferometry," Princeton University Department of Astrophysical Sciences, May 2011 (accessed at web.astro.princeton.edu) • Bemis, Ashley; Braatz, Jim; Pack, Alison, "Introduction to Radio Interferometry," National Radio Astronomy Observatory (March 16, 2015; accessed at science.nrao.edu)
PHOTO CREDITS: World History Archive/Alamy Stock Photo. *Inset:* ALMA (NRAO/ESO/NAOJ); C. Brogan, B. Saxton (NRAO/AUI/NSF). Distributed under the CC By-SA 3.0 license.

50   The Heat Shield
Freudenrich, Craig, "How Project Mercury Worked," HowStuffWorks.com (May 4, 2001) • Port, Jake, "How Do Heat Shields on Spacecraft Work?," *Cosmos* (May 4, 2016)
PHOTO CREDITS: Smithsonian National Air and Space Museum. *Inset:* NASA.

51   The Integrated Circuit
PHOTO CREDIT: NASA.

52   The Atomic Clock
Chen, Sophia, "These Super-Precise Clocks Help Weave Together Space and Time," *Wired* (May 1, 2019) • Earth & Sky • Horton, J. W., "Precision Determination of Frequency," *Proceedings of the Institute of Radio Engineers* 16(2): 137–154, 1928
PHOTO CREDIT: National Institute of Standards and Technology.

53   Space Fasteners
PHOTO CREDITS: NASA (both images).

54   The Hydrogen Line Radio Telescope
National Radio Astronomy Observatory: nrao.edu • Van de Hulst, H. C.; Muller, C. A.; Oort, J. H., "The Spiral Structure of the Outer Part of the Galactic System Derived from the Hydrogen Emission at 21 cm Wavelength," *Bulletin of the Astronomical Institutes of the Netherlands* 12: 117, 1954
PHOTO CREDITS: Photos courtesy Green Bank Observatory/GBO/AUI/NSF (left page). Benjamin Winkel & HI4PI collaboration (right page).

55   The X-Ray Imaging Telescope
Chandra X-Ray Observatory: chandra.harvard.edu
PHOTO CREDITS: NASA. *Inset:* Wikipedia/Lucie Green. Distributed under the CC BY-SA 3.0 license.

56   The Hydrogen Bomb
Atomic Heritage Foundation: atomicheritage.org • Pappas, Stephanie, "Hydrogen Bomb vs. Atomic Bomb: What's the Difference?," LiveScience (September 22, 2017) • Rathi, Akshat, "Why It's So Difficult to Build a Hydrogen Bomb," Quartz (January 7, 2016)
PHOTO CREDITS: US National Nuclear Security Administration/Nevada Site Office. *Inset:* Wikipedia/Croquant. Distributed under the CC BY-SA 3.0 license.

57   The Radioisotope Thermoelectric Generator
Jiang, Mason, "An Overview of Radioisotope Thermoelectric Generators," Stanford University Department of Physics, Winter 2013 (accessed at physics.stanford.edu) • US Department of Energy: energy.gov
PHOTO CREDITS: Department of Energy (both images).

58   The Nuclear Rocket Engine
David Darling: daviddarling.info • National Archives, Pieces of History: prologue.blogs.archives.gov • Taub, J. M., "A Review of Fuel Element Development for Nuclear Rocket Engines," Los Alamos Scientific Laboratory, 1975
PHOTO CREDITS: NASA. *Inset:* Public Domain.

59   Sputnik
PHOTO CREDIT: NASA.

60   *Vanguard 1*
Hollingham, Richard, "The World's Oldest Scientific Satellite Is Still in Orbit," *BBC News* (October 6, 2017) • Locklear, Mallory, "Vanguard I Has Spent Six Decades in Orbit, More Than Any Other Craft," Endgadget (March 16, 2018)
PHOTO CREDITS: NASA (both images).

61   *Luna 3*
Long, Tony, "Oct. 7, 1959: Luna 3's Images from the Dark Side," *Wired* (October 7, 2011) • Zarya: zarya.info
PHOTO CREDITS: NASA. *Inset:* Public Domain.

62   The Endless Loop Magnetic Tape Recorder
Engineering and Technology History • History-Computer • Museum of Magnetic Sound Recording: museumofmagneticsoundrecording.org • The National Valve Museum: r-type.org • Newville, Leslie J., "Development of the Phonograph at Alexander Graham Bell's Volta Laboratory," *Contributions from the Museum of History and Technology, United States National Museum Bulletin* 218, Paper 5: 69–79, 1959 • Stark, Kenneth W.; White, Arthur F., "Survey of Continuous-Loop Magnetic Tape Recorders Developed for Meteorological Satellites," National Aeronautics and Space Administration, 1965
PHOTO CREDITS: Wikipedia/Sanjay Acharya. Distributed under the CC BY-SA 4.0 license. *Inset:* NASA.

63   The Laser
Maiman, Theodore H., speech at press conference on July 7, 1960 (accessed at hrl.com)
PHOTO CREDITS: ESO/Gerhard Hudepohl. *Inset:* CC0.

64   Space Food
Calderone, Julia, "Astronauts Crave Spicy Food in Space—Here's Why," *Business Insider* (February 6, 2016) • Mental Floss: mentalfloss.com • Sang-Hun, Choe, "Starship Kimchi: A Bold Taste Goes Where It Has Never Gone Before," *The New York Times* (February 24, 2008)
PHOTO CREDITS: NASA (both images).

65   The Space Suit
"From Mercury to Starliner: The Evolution of the Spacesuit," *NBC News* (February 20, 2017; accessed at nbcnews.com) • Hanson, Roger, "The Armstrong Limit," Stuff (August 5, 2016) • Kerrigan, Saoirse, "The Evolution of the Spacesuit: From the Project Mercury Suit to the Aouda.X Human-Machine Interface," Interesting Engineering (May 18, 2018) • New Mexico Museum of Space History: nmspacemuseum.org • US Rocket Academy, Citizens in Space: citizensinspace.org
PHOTO CREDIT: NASA.

66   *Syncom 2* (and *3*)
John F. Kennedy Presidential Library and Museum: jfklibrary.org • "*Syncom 3* Is Launched into a Preliminary Orbit; Satellite to Be Moved to Point Over the Pacific to Relay Olympic TV From Tokyo," *The New York Times* (August 20, 1964) • Via Satellite: satellitetoday.com
PHOTO CREDITS: NASA (both images).

67   The Vidicon Camera
Drew Ex Machina: drewexmachina.com • Hungarian Intellectual Property Office: hipo.gov.hu/en • Space Loot: venusianw.tumblr.com • Teletronic: teletronic.co.uk
PHOTO CREDITS: NASA (left page). Wikipedia/Mike Peel (right page). Distributed under the CC-BY-SA-4.0 license.

68   The Space Blanket
Oetken, Nick, "The Benefits of Space Blankets in a Survival Situation," Outdoor Revival (March 23, 2018)
PHOTO CREDITS: Panther Media GmbH/Alamy Stock Photo. *Inset:* NASA.

69   The Handheld Maneuvering Unit
Drake, Nadia, "First Person to Walk Untethered in Space Gives a Final Interview," *National Geographic* (February 7, 2018) • SciHi: scihi.org
PHOTO CREDITS: NASA (both images).

70   *Apollo 1* Block I Hatch
PHOTO CREDITS: Smithsonian National Air and Space Museum (left page). NASA (right page).

71   The Interface Message Processor
Communications Museum Trust: communicationsmuseum.org.uk • Computer History Museum: computerhistory.org • History-Computer • University of California, Los Angeles, Information Studies Research Lab: islab.gseis.ucla.edu • Internet Hall of Fame: internethalloffame.org • "'Lo' and Behold: A Communication Revolution," *NPR: All Things Considered* (October 29, 2009) • World Wide Web Foundation: webfoundation.org • Zakon Group: zakon.org
PHOTO CREDIT: Wikipedia/Steve Jurvetson. Distributed under the CC BY 2.0 license.

72   The Hasselblad Camera
Hasselblad: hasselblad.com • Phillips, Henry, "Hasselblad's History in Space," Gear Patrol • Savov, Vlad, "This Is How the World's Most Covetable Cameras Get Made," The Verge (February 6, 2018)
PHOTO CREDITS: NASA (both images).

73   *Apollo 11* Moon Rocks
Lunar and Planetary Institute: lpi.usra.edu • Roberts, Sam, "How Moon Dust Languished in a Downing Street Cupboard," *The New York Times* (January 13, 2016)
PHOTO CREDIT: Wikipedia/Mitch Ames. Distributed under the CC BY-SA 4.0 international license.

74   The CCD Imager
Cakebread, Caroline, "People Will Take 1.2 Trillion Digital Photos This Year—Thanks to Smartphones," *Business Insider* (August 31, 2017) • Large Synoptic Survey Telescope: lsst.org • University of Arizona Department of Astronomy and Steward Observatory: as.arizona.edu
PHOTO CREDITS: NASA (both images).

75   The Lunar Laser Ranging RetroReflector
Lunar and Planetary Institute
PHOTO CREDIT: NASA.

76   The Apollo Lunar Television Camera
Smithsonian National Museum of Natural History: naturalhistory.si.edu • Teital, Amy Shira, "How NASA Broadcast Neil Armstrong Live from the Moon," *Popular Science* (February 5, 2016)
PHOTO CREDITS: NASA (both images).

77   The Homestake Gold Mine Neutrino Detector
APS News • Brown, Laurie M., "The Idea of the Neutrino," *Physics Today* 31(9): 23, 1978 • INSPIRE, High-Energy Physics Literature Database: inspirehep.net • Kamioka Observatory Institute for Cosmic Ray Research: http://www-sk.icrr.u-tokyo.ac.jp/index-e.html
PHOTO CREDIT: Science History Images/Alamy Stock Photo.

78   *Lunokhod 1*
Crane, Lea, "First Photo of Chinese Yutu-2 Rover Exploring Far Side of the Moon," *New Scientist* (January 3, 2019) • Zak, Anatoly, "The Day a Soviet Moon Rover Refused to Stop," *Air & Space* (January 18, 2018)
PHOTO CREDIT: SPUTNIK/Alamy Stock Photo.

79   The Skylab Exercise Bike
Pickrell, John, "Timeline: Human Evolution," *New Scientist* (September 4, 2006) • Power & Speed Training Company: powerspeed-training.com
PHOTO CREDITS: NASA (both images).

80   The Laser Geodynamics Satellite (LAGEOS)
Choi, Charles Q., "Strange But True: Earth Is Not Round," *Scientific American* (April 12, 2007) • Lynch, Peter, "That's Maths: Earth's Shape and Spin Won't Make You Thin," *Irish Times* (November 20, 2014) • Universe Today: universetoday.com
PHOTO CREDITS: NASA. *Inset:* Courtesy of the GFZ German Research Centre for Geosciences.

81   Smoot's Differential Microwave Radiometer
European Space Agency: esa.int • The Nobel Prize: nobelprize.org • Smoot, George F., "Cosmic Microwave Background Radiation Anisotropies: Their Discovery and Utilization," Nobel Lecture (December 8, 2006; accessed at nobelprize.org) • Smoot Group, Berkeley Lab: aether.lbl.gov • Theodora.com
PHOTO CREDITS: NASA (both images).

82   The Viking Remote-Controlled Sampling Arm
The Planetary Society: planetary.org
PHOTO CREDIT: NASA.

83   The "Rubber Mirror"
American Astronomical Society • European Southern Observatory • Lawrence Berkeley National Laboratory (Berkeley Lab): lbl.gov • Olivier, Scot, "A New View of the Universe," *Science & Technology Review*, July/August 1999 • Max, Claire, "Introduction to Adaptive Optics and its History," American Astronomy Society • Sanders, Robert, "Physicist Frank Crawford, Who Worked on Bubble Chambers, Supernovas and Adaptive Optics, Has Died at 79," UC Berkeley News, 2003
PHOTO CREDITS: ESO. *Inset:* ESO/P. Weilbacher (AIP).

84   The Multi-Fiber Spectrograph
Hill, J. M., "The History of Multiobject Fiber Spectroscopy," ASP Conference Series 3 (Fiber Optics in Astronomy): 77, 1988 • Ratcliffe, Martin A., *State of the Universe 2008: New Images, Discoveries, and Events*, New York: Springer, 2008 • Sloan Digital Sky Survey: sdss.jhu.edu
PHOTO CREDITS: Phil Massey, Lowell Obs./NOAO/AURA/NSF. *Inset:* ESO.

85   The Venera Landers
Teitel, Amy Shira, "Yes, We've Seen the Surface of Venus," *Popular Science* (January 6, 2015)
PHOTO CREDITS: NASA. *Inset:* Public Domain.

86   The Compromised *Challenger* O-Rings
The Rogers Commission Report (accessed at er.jsc.nasa.gov/seh/explode.html) • Than, Ker, "5 Myths About the *Challenger* Shuttle Disaster Debunked," *National Geographic* (January 22, 2016) • Wise, George, "O-Ring," *Invention & Technology* 25(3): Fall 2010
PHOTO CREDITS: NASA (both images).

87   COSTAR
Encyclopedia.com • University of Arizona Research, Discovery & Innovation: research.arizona.edu
PHOTO CREDITS: Image by Eric Long, Smithsonian National Air and Space Museum. *Inset:* NASA.

88   CMOS Sensors
B & H Foto & Electronics Corp.: bhphotovideo.com • De Moor, Piet, "CMOS, CCDs Invade Space Imagers," *EE Times* (November 26, 2013) • Pepitone, Julianne, "Chip Hall of Fame: Photobit PB-100," *IEEE Spectrum* (July 2, 2018) • Queen Elizabeth Prize for Engineering: qeprize.org
PHOTO CREDIT: Wikipeia/Weirdmeister. Distributed under the CC BY-SA 4.0 international license.

89   The Allan Hills Meteorite
Lunar and Planetary Institute • National Academies of Sciences, Engineering, Medicine: nap.edu
PHOTO CREDITS: NASA (both images).

90   *Sojourner*
PHOTO CREDITS: NASA (both images).

91   Gravity Probe B
Cho, Adrian, "At Long Last, Gravity Probe B Satellite Proves Einstein Right," *Science* (May 4, 2011) Gugliotta, Guy, "Perseverance Is Paying Off for a Test of Relativity in Space," *The New York Times* (February 16, 2009) • European Southern Observatory • Stanford University W. W. Hansen Experimental Physics Lab, Gravity Probe B: einstein.stanford.edu • Guinness World Records: guinnessworldrecords.com • Hecht, Jeff, "Gravity Probe B Scores 'F' in NASA Review," *New Scientist* (May 20, 2008) • Will, Clifford M., "Viewpoint: Finally, Results from Gravity Probe B," *Physics* 4(43), 2011
PHOTO CREDIT: NASA.

92   LIDAR
Bryan, Thomas C.; Howard, Richard T., "The Next Generation Advanced Video Guidance Sensor: Flight Heritage and Current Development," *AIP Conference Proceedings* 1103(615), 2009 • Carrington, Connie K.; Heaton, Andrew; Howard, Richard T; Pinson, Robin M., "Orbital Express Advanced Video Guidance Sensor, *IEEE Aerospace Conference Proceedings*, 2008 • Christian, John A.; Cryan, Scott, "A Survey of LIDAR Technology and its Use in Spacecraft Relative Navigation," American Institute of Aeronautics and Astronautics: Guidance, Navigation, and Control (GNC) Conference, 2013 • European Space Agency • Frey, Randy W., "LADAR Vision Technology for Automated Rendezvous and Capture," *NASA Automated Rendezvous and Capture Review*, 1991 Hillhouse, Jim, "Orion Rendezvous

Technology Launches on Next Shuttle Flight," AmericaSpace (April 5, 2010) • Molebny, Vasyl; McManamon, Paul F.; Steinvall, Ove; Kobayashi, Takao; Chen, Weibiao, "Laser Radar: Historical Prospective—from the East to the West," *Optical Engineering* 56(3), 2016 • *Selected Highlights from 25 Years of Missile Defense Technology Development & Transfer: A Technology Applications Report* (accessed at discover.dtic.mil) • Sensors Unlimited: sensorsinc.com • Space Foundation: spacefoundation.org • Szondy, David, "ESA Tests New Rendezvous System as ATV-5 Docks at Space Station," New Atlas (August 13, 2014)
PHOTO CREDITS: NASA (both images).

93   The Large Hadron Collider
Conover, Emily, "The Large Hadron Collider Is Shutting Down for 2 Years," ScienceNews (December 3, 2018) • Fermilab: fnal.gov • "Large Hadron Collider," *symmetry: dimensions of particle physics* (August 1, 2006) • Worldwide LHC Computing Grid: wlcg-public.web.cern.ch
PHOTO CREDIT: Maximilien Brice, CERN. Distributed under the CC BY-SA 3.0 license.

94   The Kepler Space Telescope
Alonso, Roi; Deeg, Hans J., "Transit Photometry as an Exoplanet Discovery Method," *Handbook of Exoplanets*, New York: Springer, 2018 • Clery, Daniel, "Kepler, NASA's Planet-Hunting Space Telescope, Is Dead," *Science* (October 30, 2018) • Gary, Dale E., "Astrophysics I: Lecture 10, Search for Extrasolar Planets" (accessed at web.njit.edu) • Juncher, Diana, "How Do Scientists Find New Planets?," ScienceNordic (January 12, 2018) • The Planetary Society • Wehner, Mike, "NASA's Kepler Just Spotted 18 New Earth-Sized Planets, but Only One Is Worth Dreaming About," BGR (May 23, 2019)
PHOTO CREDITS: NASA (left page). NASA Ames/SETI Institute/JPL-Caltech (right page).

95   *Curiosity* Rover
Kerr, Dara, "Viewers Opted for the Web Over TV to Watch Curiosity's Landing," *CNET* (August 8, 2012)
PHOTO CREDITS: NASA. *Inset:* NASA/JPL-Caltech/LANL.

96   *Mangalyaan*—the Mars Orbiter Mission
PHOTO CREDIT: Getty Images/Pallava Bagla.

97   A 3-D Printed Ratchet Wrench
SpaceX: spacex.com
PHOTO CREDIT: NASA.

98   The LIGO Gravitational-Wave Interferometer
Blair, David, "New Detections of Gravitational Waves Brings the Number to 11—so Far," The Conversation (December 3, 2018) • Brooks, Michael, "Grave Doubts Over LIGO's Discovery Of Gravitational Waves," *New Scientist* (October 31, 2018) • Event Horizon Telescope: eventhorizontelescope.org • Francis, Matthew, "The Dawn of a New Era in Science," *The Atlantic* (February 11, 2016) • Gretz, Darrell J., "Early History of Gravitational Wave Astronomy: The Weber Bar Antenna Development," Forum on the History of Physics Newsletter, Spring 2018 • LIGO Laboratory: ligo.caltech.edu • Lindley, David, "A Fleeting Detection of Gravitational Waves," *Physical Review Focus* 16(19), 2005 • O'Neill, Ian, "Gravitational Waves vs. Gravity Waves: Know the Difference!," LiveScience (February 11, 2016) • Siegfried, Tom, "Einstein's Genius Changed Science's Perception of Gravity," ScienceNews (October 4, 2015) • Woodford, Chris, "Interferometers," ExplainThatStuff! (November 5, 2018)
PHOTO CREDITS: Christian Offenberg/Alamy Stock Photo. *Inset:* Public Domain.

99   The Tesla Roadster
"NASA Budgets: US Spending on Space Travel Since 1958," *Guardian* Data Blog
PHOTO CREDIT: Public Domain.

100   The Event Horizon Telescope
European Southern Observatory
PHOTO CREDIT: EHT Collaboration.

ACKNOWLEDGMENTS

This complex book could not have been undertaken without the suggestion and support of my editor at The Experiment, Nicholas Cizek. It took many months and no fewer than four different drafts just to develop the list! Thanks also to the rest of the team at The Experiment, including Beth Bugler and Jack Dunnington for the design, Zach Pace and Pamela Schechter for managing the complex process of putting together and producing the book, Nancy Elgin and Allison Dubinsky for first-rate copyediting and fact checking, and Jennifer Hergenroeder and Ashley Yepsen for promoting my work.

I would also like to thank John Mather for his foreword. He's personally responsible for making some of the historic contributions I recount in this book, and I am delighted to have had his continuing friendship over the many years since our paths crossed in 1991 during the COBE mission.

Finally, I would like to thank my family for their support and understanding as I obsessively talked about these one hundred objects, getting lost in the minutiae while I wrote an essay, and then doing the same thing all over again, ninety-nine more times.

ABOUT THE AUTHOR

Dr. Sten Odenwald is an award-winning astrophysicist and prolific science popularizer who has been involved with science education for the COBE, IMAGE, Hinode, and InSight missions, as well as NASA's Sun-Earth Connection Education Forum. He is currently the director of citizen science for the NASA Space Science Education Consortium at the NASA Goddard Space Flight Center.